The Problem with Software

The Problem with Software

Why Smart Engineers Write Bad Code

Adam Barr

The MIT Press
Cambridge, Massachusetts
London, England

Excerpt from "Howl" by Allen Ginsberg. Copyright © 1956, 2010 Allen Ginsberg LLC, used by permission of The Wylie Agency LLC.

This book was set in ITC Stone Serif Std by Toppan Best-set Premedia Limited. Printed and bound in the United States of America.

Library of Congress Cataloging-in-Publication Data

Names: Barr, Adam, author.
Title: The problem with software : why smart engineers write bad code / Adam Barr.
Description: Cambridge, MA : The MIT Press, [2018] | Includes bibliographical
    references and index.
Identifiers: LCCN 2018013460 | ISBN 9780262038515 (hardcover : alk. paper)
Subjects: LCSH: Computer software--Development--Anecdotes. | Computer
    programmers--Anecdotes.
Classification: LCC QA76.76.D47 B373 2018 | DDC 005.3--dc23 LC record available
    at https://lccn.loc.gov/2018013460

10   9   8   7   6   5   4   3   2

To my children, Zachary, Madeline, Keenan, and Noah

Contents

Acknowledgments

A sincere thanks to the computer scientists who were kind enough to be interviewed by me, in person or via e-mail: Henry Baird, Victor Basili, Fred Brooks, Robert Harper, Donald Knuth, Adam McKay, David Parnas, Vaughan Pratt, and Ben Shneiderman. I'd also like to thank Glenn Dardick and Vincent Erickson, who both responded to my impromptu Facebook messages and filled in some great historical detail.

I would also like to express my deepest thanks to everybody who reviewed the book. My most heartfelt appreciation goes to my sister, Rebecca, who read through the book with her professional editor's eye—three times. I'm also grateful to my parents, Michael and Marcia, and my brother, Joe, who read the whole thing and gave lots of feedback, and Krishnan Ramaswami, who despite not being related to me, read and commented on every chapter. I'd also like to thank my son Zachary, who contributed valuable feedback, and Emily Papel and Carrie Olesen, who gave me early encouragement. Thanks to Bernard Pham, Kate Varni, and Joseph White as well for their comments. Bob Drews once again did an excellent job editing the final manuscript.

A big thank you to everyone at the MIT Press who worked on the book, particularly Maria Lufkin Lee, who expressed initial interest in and eventually acquired the book, and Christine Savage and Stephanie Cohen, who answered many questions, along with the anonymous referees who provided insightful comments. My appreciation also to Virginia Crossman, Cindy Milstein, and Susan Clark.

Thanks to all my colleagues in Engineering Excellence at Microsoft, especially those who worked together with me in Developer Excellence, and specifically Eric Brechner, who hired me into Engineering Excellence and was the guiding light behind much of what we did. I'll also mention Kristen Lane, who was kind enough to explain knob-and-tube wiring to me.

Thanks to the staff members of the King County Library System for allowing me to haunt their many branches while working on the book, and the baristas at various local coffee shops for the same—especially Yum-E Yogurt in Issaquah. The Living Computer Museum in Seattle let me relive a not-so-small part of my youth, which was much appreciated.

Various anonymous people on Reddit have contributed knowledge that directly or indirectly found its way into the book. I don't know who you are, and you didn't know who I was, but consider yourselves acknowledged. I'm also grateful to the maintainers of whatever Wikipedia pages I used in researching the book.

Finally, I would like to thank my wife, Maura, for putting up with the giant pile of dusty software books, and with me, as always.

Introduction

In November 1988, a computer virus attacked computers connected to the still-nascent Internet. The virus exploited a programmer error: assuming that another computer could be trusted to send the right amount of data. It was a simple mistake, and the fix was trivial, but the programming language used was vulnerable to this type of mistake, and there was not a standard methodology for detecting that sort of problem.

In April 2014, a computer virus attacked computers connected to the now-ubiquitous Internet. The virus exploited a programmer error: assuming that another computer could be trusted to send the right amount of data. It was a simple mistake, and the fix was trivial, but the programming language used was vulnerable to this type of mistake, and there was not a standard methodology for detecting that sort of problem.

Remaining stuck on "vulnerable programming language" and "no way to detect mistakes" isn't what we expect after a quarter century of progress. Other new engineering disciplines started out producing unreliable products. In the early days of aviation, people built planes in their garages, with predictable results. Now, a hundred years later, when a world without air travel is unimaginable, we have extremely reliable planes based on well-understood and agreed-on engineering standards.

Not so with writing software. Although labeled an engineering discipline, software has few of the hallmarks of engineering, where a body of knowledge is built up over time based on rigorous experimentation. Questions one would reasonably ask of an engineered product—How strong is it? How long will it last? How it might fail?—cannot be reliably answered for software, for either an individual part of a program or an entire suite of software. Professional licensing, a hallmark of most engineering disciplines, is viewed by the software industry as a potential source of lawsuits rather than an opportunity to establish standards.

The effect of this is not just user-visible bugs; it's also a lot of wasted effort and reinvention on the part of programmers, leading to frustration and software that is delayed or never ships.

If you've heard about the software industry, it might be because of the unusual way in which programmers are interviewed. Websites, books, and even weeklong training classes are devoted to preparing people for the dreaded coding interview, which is presented as an all-or-nothing chance to impress with your skills and knowledge—especially through "whiteboard coding," in which a candidate has to dash out short programs on the whiteboard. Some candidates complain that this isn't an accurate representation of what their daily job would be, and they want companies to focus on other areas of their background. What they may not realize is that there isn't much else in their background to focus on. Unlike in other engineering disciplines, having a degree in software engineering does not guarantee that you understand a known corpus of programming tools and techniques, because such a thing does not exist. You likely wrote a lot of code when you were in college, but there is no way of knowing if it was any good. So asking people to write code snippets on a whiteboard is the best way we have to evaluate people.

Consider this joke, although it's no laughing matter: What do you call the person who graduated last in their medical school class? The answer is "doctor"—because graduating from medical school and completing your residency implies that you have learned what is needed to be one. I have asked doctors how they were interviewed when they were hired. They say that they were never asked specific medical questions or to perform simple medical procedures; instead, the talk was about how they speak to patients, how they feel about new medicines, and that sort of thing—because it is understood that they know the basics of medicine. Computer science graduates can make no such universal claim.

Back in November 1990, Mary Shaw of Carnegie Mellon University wrote an article for *IEEE Software* magazine titled "Prospects for an Engineering Discipline of Software." Shaw explains that "engineering relies on codifying scientific knowledge about a technological problem domain in a form that is directly useful to the practitioner, thereby providing answers for questions that commonly occur in practice. Engineers of ordinary talent can then apply this knowledge to solve problems far faster than they otherwise could. In this way, engineering shares prior solutions rather than relying always on virtuoso problem solving." She compares software to civil engineering, pointing out, "Although large civil structures have been built since before recorded history, only in the last few centuries has their design and construction been based on theoretical understanding rather than on intuition and accumulated experience."[1] As I leaf through the publications catalog of the American Society of Civil Engineers, full of intriguing titles such as *Water Pipeline Condition Assessment* and *Cold Regions Pavement Engineering*, I can appreciate how much theoretical understanding there is in other engineering disciplines.

Looking back at the history of engineering in various forms, Shaw writes, "Engineering practice emerges from commercial practice by exploiting the results of a companion science. The scientific results must be mature and rich enough to model practical problems. They must also be organized in a form that is useful to practitioners."[2] Yet in the years since her article appeared, the software engineering community has made little progress in building up the scientific results needed to support a true engineering discipline; it is still stuck in the "intuition and accumulated experience" phase. At the same time, software has become critically important to modern life; people assume it is much more reliable than the underlying engineering methodologies can guarantee.

Shaw ends her article with the following: "Good science depends on strong interactions between researchers and practitioners. However, cultural differences, lack of access to large, complex systems, and the sheer difficulty of understanding those systems have interfered with the communication that supports these interactions. Similarly, the adoption of results from the research community has been impeded by poor understanding of how to turn a research result into a useful element of a production environment. ... Simply put, an engineering basis for software will evolve faster if constructive interaction between research and production communities can be nurtured."[3]

At the 2013 Systems, Programming, Languages, and Applications: Software for Humanity (SPLASH) conference, sponsored by the Association for Computing Machinery (ACM), a professional organization, a programmer named Greg Wilson gave a keynote talk titled "Two Solitudes" about this divergence between academia and industry in the world of software. After working as a programmer for a while, Wilson discovered the landmark book *Code Complete*, one of the first to attempt to lay out a practice of software engineering, and one of the rare software books that references research studies on software practices. Wilson realized he had been previously unaware of all this; as he said in the talk, "How come I didn't know we knew stuff about things?"[4] Then he realized that none of his coworkers did either, and furthermore, they were happy in their ignorance and had no desire to learn more. He also commented, "Less than 20% of the people who attend the International Conference on Software Engineering come from industry, and most of those work in labs like Microsoft Research. Conversely, only a handful of grad students and one or two adventurous faculty attend big industrial conferences like the annual Agile get-together."[5]

The hand-wringing over software engineering has been going on since the term was invented fifty years ago. This book won't propose a solution, although it includes suggestions at the end, but it will attempt to provide a guided tour of the path that the software industry has taken from its early days to the present.

With a couple of exceptions, the chapters are arranged in chronological order, roughly paralleling my own experience as a programmer, starting around 1980. The book is not attempting to be a complete history of the software industry; rather, it digs into specific moments that are especially important and representative. Those moments involve a succession of ideas that were touted as the one single solution to all the problems facing programmers, before inevitably falling back to earth and being superseded by the next big thing. At the same time, the gap between academia and industry has continued to widen, so that each new idea becomes less moored in research, and software drifts further from, not closer to, the engineering basis that Shaw was hoping for.

Fundamentally, the book is about a question that I have often asked myself: Is software development really hard, or are software developers not that good at it?

Spoiler alert for technophobes: there is some code in this book. Do not be dismayed. It is impossible to understand the software industry without understanding what programmers are thinking about, and it's impossible to understand what programmers are thinking about without digging into the actual code they write. The difference between good and bad software can be a single line of code—a seemingly inconsequential choice made by a programmer. To understand some of the problems with software, you need to understand enough about code to appreciate that difference, and why programmers write the bad line of code instead of the good line.

So please read the code! Thank you.

1 Early Days

I am holding in my hand the 1982 Radio Shack computer catalog, unaccountably saved from my high school years. Sold under the TRS-80 brand, the products look familiar: computers with displays and keyboards, printers and hard drives as well as games and productivity software. The prices are reasonable, if a little high by today's standards: $800 for a basic desktop computer, $1,100 for one with extra memory, and a more powerful business system for $2,300. There are no laptops (although there is a "pocket computer"), and touch screens are far in the future, but everything else is recognizable.[1]

It's the details that are jarring. Compared to current hardware, the capacity of these computers is microscopic. The $800 TRS-80 Model III, the entry-level model, has 4 kilobytes of memory—a mere 4,096 bytes. From the perspective of today, when a low-end computer might have 4 gigabytes of memory, it is hard to conceive that one with 4 kilobytes—one-millionth as much—could do anything useful at all.

The storage difference is similar. A hard drive selling for $5,000 holds 8.4 megabytes; today you can buy an 8-terabyte hard drive—larger, again, by a factor of a million—for less than $300. The floppy disks hold 170 kilobytes each (once you pay almost $1,000 more for a floppy disk drive and disk operating system software); today's USB sticks hold a million or more times as much. Shrinking a modern storage device to the size of what was available back then would be the equivalent of reducing this entire book to half of one letter. A car that moved at one-millionth of the speed of today's cars would move on the order of inches per hour; such a car would be considered, by normal human observation, to be standing still.

The year 1982 is also a milestone in my personal history: my family acquired our first home computer, an original IBM Personal Computer running IBM PC DOS (Disk Operating System) version 1.00.

Before the appearance of the IBM PC in late 1981, the personal computer industry was split between incompatible computers from Radio Shack, Apple, and Commodore. Anybody wanting to sell software for all three had, essentially, to write it three times. The IBM PC introduced a fourth platform, but for a variety of reasons—the facts that IBM was the most well-known computer company in the world whose name "legitimized" the personal computer industry, the IBM PC had a more expandable hardware design than its competitors, Microsoft sold a version of DOS to other hardware companies, and a company called Phoenix Technologies wrote a lawsuit-proof copy of the low-level software that IBM included on the computer, or possibly because of serendipity and other factors, and all these in unclear proportions—the IBM PC soon became the standard for all personal computers, supported by a robust marketplace of companies selling PC-compatible computers, consigning the other three to the recycling bin of history. With a standard platform to target, the personal computer software industry began its rapid growth. And if these slow, underpowered computers seemed like barren soil in which to grow such an enterprise, in terms of the programmer's experience they were light-years ahead of what they replaced.

As a small child I had limited exposure to actual computers due to the fact that much of the technology would not fit through our front door. I prepared for my future career by playing with Lego bricks—a common thread in the life story of programmers my age. In the mid-1970s, computers existed in two forms: mainframe computers, the big ones you see in old movies, which were used for weather forecasting and whatnot, and tended to be owned by large companies, governments, and universities; and minicomputers, which were smaller and more self-contained, and used by businesses for tasks such as running their payroll. The notion that somebody might have a computer in their house was seen as both frivolous and ridiculous; computers were for dull, important things, and in any case, where would you put it? When Microsoft was founded in 1976, the corporate vision of "a computer on every desk and in every home" seemed, well, a vision.

The year 1977 saw the introduction of the first three broadly successful personal computers, the Commodore PET, Tandy TRS-80, and Apple II, along with the Atari 2600 game system (then, as now, game systems were also computers, with a different user interface and nominal purpose).

Meanwhile, at a far remove from those scrappy upstarts, somebody in the math department at McGill University, where my father was a professor, convinced the department to purchase a minicomputer made by a company called Wang, to the tune of $20,000 plus $2,000 for the annual service contract. While mainframes tended to be the size of industrial refrigerators and installed behind glass in climate-controlled rooms, minicomputers were the size of smaller appliances and could be installed anywhere. The entire category of minicomputers would eventually be obliterated by the descendants of the original personal computers (which at the time were known by the vaguely insulting term *microcomputer*), and within fifteen years Wang Laboratories would suffer a precipitous *Innovator's Dilemma* drop—to borrow the title from Clayton Christensen's book—from thirty thousand employees to bankruptcy, but at the time Wang computers were considered quite capable.[2]

My childhood home did not participate in the first round of personal computer adoption—no Apple, Commodore, or TRS-80, or even a gaming system in our living room—but my father would occasionally bring me to the McGill University math department on a Saturday, where I could while away the afternoon on the Wang minicomputer playing games like *Bowling*, *Football*, and *Star Trek*—incredibly unsophisticated versions, with "graphics" lovingly rendered in text (the football game constructed the field markings out of dashes, plus signs, and capital *I*'s, and at halftime indicated the presence of the band by moving the letters *B-A-N-D* around the screen). I did this perhaps three times a year and can still remember the feeling of anticipation as a visit approached. As I write this I am on an airplane with two of my children. Looking over I see that—well, as it happens, they are both reading books right now, but half an hour ago they were playing games on handheld devices. The difference between my access to computer games and theirs, to say nothing of the quality of the games, is mind-blowing: as if I had observed evolution from trilobite to *Tyrannosaurus rex* in a single generation.

This introduction to playing computer games was roughly aligned with the invasion of the video game *Pong* into homes across the land, so I was not unique in spending time squinting at poorly drawn electronic entertainment, but my introduction to writing software came early for that era. The first software that I ever wrote was not on a computer. Hewlett-Packard had a line of calculators that could execute programs written by the user (as the manual for one of them states, "Because of their advanced capabilities, these calculators can even be called personal computing systems").[3] In the late 1970s, my father owned several HP calculators, including one with a built-in thermal printer (the HP-19C). The calculators were focused on mathematical operations, such as calculating mortgage payments, but I had no need for that. I was interested in programming for programming's sake, so I wrote programs that were useless in the real world, such as one to print out prime numbers (I've never found myself with the need for a list of prime numbers, but I've written the solution in several programming languages).

Before I talk more about programming, I should explain a bit about how computers run programs. The processor, the chip inside a computer, has a set of *registers* that can each store a single number. The processor can perform operations on the numbers in these registers—addition, multiplication, and so on—and can also perform tests on them—to see if they equal to a certain value or if one is larger than the other, say—and jump to a different location in the program if the test is true. Finally, since the set of registers is small (eight to thirty-two registers is typical), the processor can move values from a register to the main computer memory, and from the main computer memory to the register, so that they can be preserved and brought back to a register when needed.

That's basically what the processor can do: operations on registers, comparisons between those registers, jumps based on those comparisons, and moving data back and forth between registers and memory. There's a bit more (such as the ability to perform mathematical operations directly on data in memory rather than having to move it into a register first), but essentially everything is built up from combinations of these *instructions*, as single processor operations are known.

The program that a computer is running is a series of instructions. For example, on the Intel processors used in many personal computers today there is an instruction called ADD that can add two registers. To add the EAX register to the ECX one (and store the new total in ECX), the instruction in human-readable form is written as:

```
ADD ECX, EAX
```

but for the machine, it is actually this sequence of bits:[4]

```
0 0 0 0 0 0 0 1 1 1 0 0 0 0 0 1
```

which is a "series of 0s and 1s" as used in the sentence "computer processors interpret a series of 0s and 1s."[5]

The sequence of 0s and 1s is known as *machine language*, and the human-readable form

```
ADD ECX, EAX
```

is known as *assembly language*; a program called an *assembler* can convert assembly language into machine language so that the computer can execute the program.

Getting back to programming on the HP calculators, this was an assembly language experience: my programs had to move data back and forth from memory to the location where the processor could operate on it (which was technically a *stack*, not registers, but the effect was similar to programming on a processor that had only two registers).[6] There are those who say that every programmer should learn assembly language first so that they know what is going on under the covers, but in truth these HP calculator programs, while notable for their time, are prone to simple mistakes and hard to read.

Toward the end of 1980, my father lugged home a device known as a *terminal* that allowed us to connect to the mainframe computer at McGill. The setup was archaic by today's standards—it was archaic by the standards of a couple years later—but this is understandable given that it was the approximate midpoint between the dawn of computing and now. The terminal was of a type known as a *teleprinter* or *teletypewriter*, also called a *line terminal*. It had a keyboard but no screen, only a printer, so it looked like a large typewriter that consumed fanfold paper (that continuous folded paper with removable holes along the side, now only seen in old movies and car rental agencies). The device sat there in an unanimated state until we dialed the phone number for a mainframe computer at McGill and placed the phone handset into an acoustic coupler.

The acoustic coupler, which looked like two oversize headphone earpieces mounted on a box (Radio Shack was selling one for $240 in its 1982 catalog),[7] cradled the handset snugly enough that it could transmit beeps and boops to the microphone and hear the same coming from the speaker. These sounds were used to exchange data with the McGill computer, which housed all the actual processing power; if you imagine a modern desktop computer with a keyboard, display, and system unit, the keyboard and display were in our house and the system unit was at McGill, attached not by a short cable but by the telephone connection. And a rather slow connection it was, with the speed being 300 baud, or about thirty characters per second, roughly (you guessed it) one-millionth of the bandwidth of a typical broadband connection today. The only good thing that I can say about the slow connection is that it made the slow printing speed of the terminal a nonissue; since a full line of eighty characters of text took about three seconds to transmit, the printer had no trouble keeping up with it.

On this system, sitting in my parents' bedroom, I learned a programming language called WATFIV, a version of the grizzled computer language Fortran.[8] Compared to assembly language, this was a *higher-level language*, providing useful abstractions such as the ability to give names to storage locations (known as *variables*) rather than requiring the use of processor register names such as EAX and ECX. A program called a *compiler* converted the higher-level language into machine language (conceptually the compiler converted it into assembly language that was then converted to machine language, but typically it blasted out 0s and 1s directly).

Since WATFIV was designed to run on mainframe computers that were often accessed from coal-powered line terminals like the one I was using, it had limited output functionality. There was no support for drawing graphics on a screen or playing sound out of a speaker, since terminals frequently had neither a screen nor a speaker. The input/output functionality of a WATFIV program was limited to reading in and printing out lines of text.

My formal training as a Fortran programmer consisted of the book *Fortran IV with WATFOR and WATFIV*, which I would eyeball during slow moments in high school history class. The book taught me the basic syntax of the language, but didn't explain how to write a program that accomplished something useful, any more than knowing the syntax of the English language allows you to put together a sales pitch or marriage proposal.

Here is an example of a WATFIV program from the book—and if ever a programming language was well suited to being printed in uppercase in a fixed-width font (which I will use for all program fragments in the book), it's a Fortran dialect (in fact, the entire book was printed in a fixed-width font):[9]

```
C EXAMPLE 6.3 - SUMMING NUMBERS
      INTEGER X,SUM
      SUM=0
    2 READ,X
```

```
      IF(X.EQ.0)GO TO 117
      SUM=SUM+X
      GO TO 2
117   PRINT,SUM
      STOP
      END
```

This program is not particularly hard to read; if you instinctively skipped over it, I encourage you to go back. The set of steps in a program is often called *code*, but not in the sense of an impenetrable mystery as in *The Da Vinci Code*; it's just a series of instructions that follow certain rules.

The program declares two variables of type INTEGER (meaning a number) named X and SUM, initializes SUM to 0, reads in a value and stores it in X, and checks if X is 0. If X is 0, then it prints out the total and stops; otherwise it adds the value of X to a running total stored in SUM and goes back to read another value.

The READ,X and PRINT,SUM lines refer to what is known as an *API*, which stands for application programming interface. APIs are functionality provided to programs in order to accomplish certain tasks—in this case, reading a value and displaying a value, respectively. We say that code *calls* an API—that is, it tells the compiler that it wants to jump to that API to perform an operation. The APIs take *parameters*, which are *passed* to the API, providing more detailed information. In this code, READ is passed the parameter X, telling it what variable to read the value into, and PRINT is passed the arguments SUM, telling it what variable to print.

The numbers to the left of the code—2 and 117—are optional line numbers, which are needed as targets of the GO TO instructions; GO TO 2 means "jump to the line numbered 2 and continue execution from that point." Unlike in most modern languages, the spacing at the beginning of each line matters. Per Fortran's rules, the line numbers appear in columns 1 through 5, and the actual program code starts in column 7. The first line, starting with a C in column 1, is a comment, which is ignored by the compiler.

As for the data that will be read when this program calls the READ API, understand that Fortran and its variants were designed to read programs off punch cards (one punch card per line of the program), run them, and print out the results. The data would normally follow after the program, on another punch card (there had to be a special card containing only the text $ENTRY between the program and data). The READ API was defined to read from the next punch card in the pile (the version of WATFIV running on the McGill computer had been enhanced to allow you to store the program and data on the mainframe computer disk drive rather than requiring that it be reread each time from cards, and could also read input data typed at the keyboard, if you wished).[10] This particular program expected there to be a final card with the value 0, which would indicate the end of the data.

These were typical of the Fortran programs that people wrote—simple solutions to problems such as calculating the semester grades of students from their individual test scores (one test score per punch card, natch). In many cases, the people who wrote programs were the same people who used those programs, because the programs they wrote were specific to their exact situation.

In this day of Xbox gamers with their own reality shows, it's hard to remember that not that long ago, playing computer games was still viewed as a geeky hobby. Many people who played computer games also dabbled in writing computer games, which was unquestionably geeky. You may not remember a British pop group called Sigue Sigue Sputnik, a one-hit wonder whose object of wonderment was the song "Love Missile F1-11"; what amazed me in 1986, looking at the back of the LP at our local Sam the Record Man store, was that these people, whose clothing and hairstyle unquestionably marked them as cool, forward-thinking types, listed "video games" as one of their hobbies—a signature moment for me. I am still slightly surprised when a student today tells me that they became interested in programming not because of the sheer thrill of 1s and 0s but instead because they enjoy using software and thought it might be interesting to learn how to write it. Nothing about Lego?

As with the HP calculators, I muddled through learning WATFIV without any particular goal in mind. I was writing simple programs; sorting a list of numbers (which in programming terms is known as an *array* of numbers) from largest to smallest was a typical example, or if I felt like an interactive experience, there was the old "the computer is thinking of a number; you guess, and I will respond with 'higher' or 'lower'" game. In order to spend the time puzzling through how to make these things work on your own, for such a meager result, you had to have a personality that viewed solving the problem as valuable in and of itself—that understood that the journey, in this situation, *was* the reward. At the same time, there's not much opportunity to make a mistake that can't be fixed by trial and error. You would spend more time discovering that a line of Fortran code mistakenly started in column 6 instead of 7 than you would in dealing with bugs in the logic of your program.

For that matter, playing with Lego as a kid probably had been decent preparation. The pieces combining into a larger object, detailed directions that had to be followed precisely, and little "click" when you connect them together are all quite similar to writing short programs.

The IBM PC that we acquired in 1982 was a much more compelling software platform than the McGill mainframe. Beyond the convenience of not having to connect remotely via a modem (after waiting until nobody else in the house was using the phone), the IBM PC could display graphics and play sounds.[11] Programmers took advantage of this to write word processing programs that displayed accurate formatting on the screen, spreadsheets that recalculated on the fly, and other marvels of the age. No more typing in text commands to a football game and then watching it respond; no more band represented by the letters *B-A-N-D* moving back and forth. Now you could write games (what I cared about) that were actually interactive!

Even better, the BASIC language included with the computer, since it was customized to run on the IBM PC, supported all this hardware. In fact, the IBM PC BASIC had advanced features that I have not seen on any other system. There was an API called PLAY, which you could feed a "tune definition language" and it would play the notes; the following

```
PLAY "L8 GFE-FGGG P8 FFF4 GB-B-4 GFE-FGGG GFFGFE-"
```

would play "Mary Had a Little Lamb."[12] And the DRAW API supported a "graphics definition language," so

```
DRAW "M100,100 R20D20L20U20 E10F10"
```

would draw a box with a triangle on top.[13]

Suddenly I had moved from a fairly difficult programming environment on the McGill mainframe, with limited commands available, and that was only really useful for small "learning to program" examples that I wrote quickly and forgot about, to a rich environment with programs that genuinely wanted to run and keep running, expand over time, and possibly even show to other people.

A cottage industry of books and magazines sprang up to serve the IBM PC community, but my main source of knowledge on BASIC was the printed manual that came with the computer. Once again, I learned by figuring it out myself—which I am pointing out not to highlight my own skill but instead to emphasize that reading a reference manual was then the standard way that people learned to program. At the same time, I was also writing BASIC code on a friend's Apple II clone, learning the details in the same way (the Apple II BASIC also had graphics and sound support, but the specifics of how they were done in BASIC differed between the PC and Apple II, and for that matter, among the BASICs included with most of the personal computers of the day).

I was able to blunder along, teaching myself enough IBM PC BASIC to crank out inferior clones of arcade games such as *Pac-Man* and *Q*bert*. My crowning achievement in high school may have been writing a program that arranged the names of every member of my senior class into a giant *84*, the year I graduated, with said design then being printed on a sweatshirt. Unfortunately, there were several factors that prevented me from learning the proper way to tackle larger programs, especially of the sort I would write in my professional career as a programmer.

The first problem was that whatever knowledge was out there about how to write "good" programs, it never made it into my consciousness. The BASIC manual had short snippets of code, each of which addressed the proper syntax and usage of a single part of the language—the language keywords and API that you would use to write BASIC programs. While there may have been clarification of what different parts of the sample were trying to accomplish, there was never any discussion of *why* a particular section of code had been written in a certain way. The code samples worked, and that was enough. The rest was up to you.

When you are dealing with small samples of code, which exist only to demonstrate calling one API, you don't spend much time worrying about readability or clarity. This came through in various ways. For example, naming a variable I or J isn't particularly helpful in a long program, where you want something more descriptive. But people often used single-letter variable names in these samples, and without much thought on the matter, this style would then be adopted in actual code. Most programmers wrote code for their own use and kept a mental image in their mind of how it worked, so they didn't concern themselves much with how readable it was by other people; I certainly didn't.

The second problem was the limited amount of memory in the computer.

The IBM PC came with three versions of BASIC in order to accommodate the resource limitations—it was possible to buy an IBM PC with as little as 16 kilobytes of memory. On such a machine you could only run Cassette BASIC, which was included with the computer, burned into a 32-kilobyte ROM (read-only memory) chip, which meant it didn't take up any space in the 16 kilobytes of memory. It was called Cassette BASIC because you could order an IBM PC without disk drives, using a cassette tape for storage. In this situation, the computer would boot directly into Cassette BASIC; there was no DOS because there was no disk to operate (or to load DOS itself from).

I assume that a few people did order a cassette-tape-only IBM PC (I vaguely recall a letter to the long-defunct magazine *SofTalk for the IBM PC* discussing exactly this), but the vast majority ordered it with floppy disks, which meant you were running DOS and therefore could launch either Disk BASIC or Advanced BASIC, known respectively as BASIC and BASICA, after the DOS commands you typed to start them.

Disk BASIC was a superset of Cassette BASIC, adding support for reading and writing files to disk as well as communicating over a modem. Advanced BASIC included that plus the advanced graphics and sound APIs like DRAW and PLAY. Disk BASIC required 32 kilobytes of memory, while Advanced BASIC needed a whopping 48 kilobytes. Keep in mind that memory had to accommodate the BASIC interpreter itself (although the more advanced BASICs did rely on some of the Cassette BASIC code stored in the separate ROM), the code for your program, and whatever memory the program itself used to store data in variables.

Given this, it becomes more understandable that you would give all your variables one-letter names because that used the least amount of memory to store the code that used those variables. BASIC allowed you to add comments in your code to explain what it was doing by preceding them with the REM statement or a ' (single-quote) character, but comments were also viewed as a space-hogging luxury. And a 16-kilobyte IBM PC wasn't even the most limited environment in which you could run BASIC; recall that Radio Shack was selling a TRS-80 computer with 4 kilobytes of memory in its lowest configuration, running a BASIC so minimal that it barely even supported character strings—variables that held sequences of text (*string* is used here as in "string of letters," not the thing you tie objects together with). Your program could have exactly two string variables in your code, named A\$ and B\$.[14]

In fact, to save memory, some of those early BASICs (although not the ones included with the IBM PC) allowed you to leave out the space characters. A *loop*, a standard programming construct that allows statements to be repeated, is normally written in BASIC like this (BASIC programs back then required a line number on every line):

```
10 FOR J = 1 TO 10
20   PRINT J
30 NEXT
```

but you can write it in one line instead:

```
10 FOR J = 1 TO 10: PRINT J: NEXT
```

which could then be jammed together as

```
10 FORJ=1TO10:PRINTJ:NEXT
```

and that would be considered completely normal and arguably clever.

These memory limits were a step backward from the mainframe days. Those mainframe computers didn't have a lot of memory either (I don't know the specific details of the McGill computer, but a typical mainframe in the 1970s had between 256 kilobytes and 1 megabyte of memory), and it was shared between all users connected at any one time, but the operating system used a technique called virtual memory that let the disk drives function as extra memory. In such an environment, shaving bytes off your code was much less important, and programmers could splurge on the occasional comment line.

The third problem that prevented me from learning proper coding techniques on an IBM PC was that at the time, the BASIC language made it difficult to write large programs. Such programs are made up of layers and layers of code, often written by different people, connected together via APIs. Code you write is typically calling an API provided by a lower layer but also supplying an API for higher layers to call. While BASIC did allow your code to call the APIs offered by BASIC itself, such as DRAW and PLAY, there was no way for your code to provide its own named APIs to other code.

BASIC did have *subroutines*, which allowed code to jump somewhere else in the program and then return back to the calling code. These are conceptually an API, yet they were referenced not by name but rather by line number (similar to the way that GO TO statements worked in Fortran or, for that matter, BASIC). Furthermore, subroutines did not support parameters; they referenced variables directly. You called a subroutine using the GOSUB (sometimes spelled GO SUB) statement, specifying a line number as the target. Consider this example (modified slightly for clarity) from the book *Structured BASIC*, from 1983:[15]

```
100 READ A
110 GO SUB 700
120 PRINT S
130 GO TO 999
700 S = 0                    ! START OF SUBROUTINE
710 FOR I = 0 TO A
720 S = S + I
730 NEXT I
740 RETURN                   !!! END OF SUBROUTINE
999 END
```

On line 120, the program is calling the subroutine starting on line 700, which sums up the numbers from 0 to A and stores the result in S. Conceptually the variable A is a parameter to that subroutine, because the subroutine uses A in its calculations, and the variable S is the *return value* from the subroutine, because setting S is the net effect of the subroutine. But those variables have no particular connection to that subroutine except in how they were used. In software parlance all variables in BASIC were *global variables*, meaning that any code could access them, whether in the main program or a subroutine. The caller of the subroutine on line 700 has to "just know" to put a certain value in A before calling the subroutine and also "just know" that the value that the subroutine calculates will be stored in S when it returns. Not to mention, it has to "just know" that the subroutine that adds the numbers from 0 to A starts at line 700, and not accidentally GO SUB to line 690 or 710.[16] The comments on lines 700 and 740, starting with a !, indicate to the reader that they are the start and end of a subroutine, but the compiler ignores comments and has no knowledge that line 700 is supposed to be the start of a subroutine.

This may seem like a minor detail, but in practice it makes calling other people's code quite tricky (of course, I was programming by myself so there was no other programmer—another factor that prevented me from learning "real" software development). Imagine copying code from another programmer and trying to include it in your own BASIC program. In addition to needing to know the exact line numbers to call subroutines as well as the exact variable names to use to pass in and return values, you have no guarantee that the line numbers and variable names used inside another person's code won't be the same as the ones you've used, which would cause a conflict. BASIC requires line numbers partly because it was designed to work on an interactive system that might not have any kind of text editor available; if you wanted to insert a line of code between two existing lines, you chose a line number that was numerically between the numbers of those two lines. And if you chose a line number that was already in use, then BASIC replaced the old line, which meant that loading somebody else's code would replace part of your code if the line numbers conflicted.[17] Essentially you were restricted to writing programs that depended only on the built-in API available in BASIC, or if they did load in other snippets of code, they were two parts of the same program that had been carefully crafted to not overlap their line numbers, as opposed to the much more generic "call an API implementation that somebody else wrote" that defines modern layered programs.

In addition, as a BASIC programmer, you got no practice in defining a clean API interface for others to call. These connections are fragile points in a program because of the potential for misunderstanding at the API boundary, whereby the caller of an API may not realize exactly what the API does, especially if it was written at a different time by a different person (and in large programs, the *source code*—the uncompiled original version—that implements the lower API is frequently not available to look at). In BASIC, you didn't have to ponder what was a clean set of variables to pass to a subroutine or return from it. Any subroutine could read or set any variable in the program. Even the subject of API naming didn't come up because only the line number identified the subroutines.

And even if you were trying to maintain a self-contained program that did not depend on any other code, you still wound up with code that was hard to read. MS-DOS, the operating system for the IBM PC, included a few sample BASIC programs to demonstrate the functionality of the language and give people something to do while looking at the computer in a store. I remember one that showed a chart of the colors (all sixteen of them) available on the IBM PC, and another that sketched the "skyline" of a "city" by drawing an endless succession of randomly sized rectangles on the screen—a pretty good effect for what must have been, at its heart, about ten lines of BASIC code, centered on an API called LINE, which would draw filled-in rectangles if you asked it properly.[18]

About half the sample programs were written by IBM, and the rest by Microsoft. According to Glenn Dardick, who supplied IBM's contribution, he wrote most of his in a couple days while recuperating from an operation. His magnum opus was the *Music* program, which would play eleven different songs, including "Pop! Goes the Weasel," "Yankee Doodle Dandy," the "Blue Danube Waltz" by Johann Strauss, and "Symphony #40" by Wolfgang Amadeus Mozart, all while a musical note followed along on a rendering of a piano keyboard. Dardick enlisted a family friend, a staff conductor at the Metropolitan Opera in New York named Richard Woitach, to help him with the music.[19]

The most famous, or infamous, of these BASIC samples was called *Donkey*, or DONKEY.BAS, since that was the name of the file containing the BASIC source code. *Donkey* was a video game with extremely simple gameplay. The screen displayed a two-lane road running from the top to bottom of the screen; near the bottom, there was a race car, which could be in either the left or right lane. A donkey would appear in one of the lanes, moving down from the top of the screen, and you hit the space bar to move the car to the other lane to avoid hitting the donkey. Once one donkey was avoided, another appeared at the top of the screen, and the timeless battle was renewed. That's the game! Videos of it in action do exist on the Internet, although don't construe this as advice to watch them. In its defense, it was written during a late-night exercise (with Bill Gates as a coauthor) to show off the power of Advanced BASIC.[20] It used both the DRAW and PLAY APIs as well as demonstrated the basics of how to write an interactive game, which was very instructive for somebody like me who was used to the noninteractive line terminal experience of a mainframe computer.

I will readily admit that I played *Donkey* a few times as an actual video game rather than merely to marvel at its simplicity. And one of my children, while reading a draft of this chapter, dug up a playable version of *Donkey* and found himself briefly captivated, especially since each successfully avoided donkey moves the race car closer to the top of the screen—a nuance I had forgotten that ratcheted up the excitement. The mobile game *Flappy Bird* became a sensation in 2014 with scarcely more compelling gameplay or more complicated controls than *Donkey* had back in 1981.

One benefit of *Donkey*'s notoriety is that the source code survives to this day, so we can see what an 8-bit era BASIC program looks like.[21] It's only 131 lines long, of which the first 45 are spent printing an introductory message and making sure you have the right hardware. You can follow the core game logic pretty well, but then suddenly you hit these two lines (in IBM PC BASIC, GOSUB was written as one word):[22]

```
1480 GOSUB 1940
1490 GOSUB 1780
```

and you have no context for what those mean—what the subroutines do, what variables they depend on being set before being called, and what variables they modify while running. At least in Fortran, the code would look something like

```
CALL LOADDONKEYIMAGE(DNK)
CALL LOADCARIMAGE(CAR)
```

which provides a hint of their purpose—load the images of the donkey and the car into the variables DNK and CAR, so they can be more easily drawn on the screen later (which is what the BASIC subroutines at lines 1940 and 1780 do). Or failing that, the authors of *Donkey* could have added comments to the lines of BASIC code:

```
1480 GOSUB 1940      ' LOAD THE DONKEY IMAGE INTO DNK
1490 GOSUB 1780      ' LOAD THE CAR IMAGE INTO CAR
```

and then the code at line 1940, instead of jumping right into this sequence to draw the donkey,[23]

```
1940 CLS
1950 DRAW "S08"
1960 DRAW "BM14,18"
1970 DRAW "M+2,-4R8M+1,-1U1M+1,+1M+2,-1"
```

could instead have started with a comment line to indicate what it was doing, maybe like this:

```
1940 REM SUBROUTINE TO DRAW DONKEY AND LOAD IT INTO DNK
```

Meanwhile DNK and CAR, at three letters each, are tied for being the longest and most descriptive variables used in the whole program; other variables include Q, D1, D2, C1, C2, and B. Note that BASIC allowed variable names to be up to forty characters long.

Another line of code in DONKEY.BAS is this one:[24]

```
1750 IF CX=DX AND Y+25>=CY THEN 2060
```

This is the collision test between the car and donkey; CX and CY are the screen coordinates of the car, while DX and Y (not DY, for no obvious reason) are the screen coordinates of the donkey. The math for these things is always a bit hard to puzzle through, so the complexity of the IF statement is expected (it's saying, if the on-screen x-coordinates are the same, meaning the car and donkey are in the same lane, and the on-screen y-coordinate of the donkey is within twenty-five of the car, then the bottom edge of the donkey is overlapping the top edge of the car, and we have a collision). The reason the code is confusing is because what the code does on collision (when the IF test is true) is jump to line 2060 (that is what THEN 2060 does) as opposed to calling an API with a name like SHOWEXPLOSION. And again, there are no comments to explain any of this. If you go to line 2060 to try to figure out what is going on, you see this:[25]

```
2060 SD=SD+1:LOCATE 14,6:PRINT "BOOM!"
```

First 1 is added to the variable SD, which isn't that informative (SD holds the number of times the donkey hit the car, for those following along at home, so it looks like it's an abbreviation of "score—donkey"), then the LOCATE statement mysteriously moves the cursor to row 14, column 6, but then you see that it prints the word "BOOM!" and you can probably guess, especially if you've played the game, that this is the collision code. But having to jump back and forth, keep all this in your mind, and depend on recognizable PRINT statements to figure out what you are looking at makes it hard to read the code.

You can imagine an alternate world in which the BASIC samples that came with MS-DOS had good variable names and helpful comments, and people took advantage of this to expand the games. I could have written SUPERDONKEY, with three lanes and two donkeys on the screen at once, except that due to the filename limitations in MS-DOS, it would have been called SUPRDONK.BAS. Still, maybe this would have incubated an appreciation of the benefits of writing code that others can understand, which we (*we* meaning "the assembled group of people that was inspired by IBM PC BASIC to want to go work at Microsoft," of which I am a member) would have then carried off to our jobs, and who knows how software engineering would have developed. But the code was hard to read, and none of that happened. Nowadays, companies that are preparing to open source their code so that the public can see it may be stressed about people criticizing their variables names or code layout, but apparently no such concerns existed back then; the only concession in the BASIC samples was a three-line IBM copyright message at the top of all of them.

Another source of BASIC code was books with titles like *BASIC Computer Games* and *More BASIC Computer Games*—two collections curated by David Ahl, founder and publisher of the early magazine *Creative Computing*. The books consisted of source code for a variety of games; the only way to play these games was to type the code in yourself, hopefully without making any mistakes (possibly, if a friend had typed it in already, they could copy it to a floppy disk or cassette tape for you). This was actually helpful in learning the language, since typing in code gave you the opportunity to think about how it worked.

Every computer had its own dialect of BASIC, partly to handle specific features of the computer, and partly because BASIC had never been standardized.[26] The net effect of this was that programs in the book would never do anything as computer specific as graphics or real-time play; they were all text based, and relied on the user typing commands and hitting "Enter" to interact with them, with lines of output being displayed one at a time. This made them suitable for almost any BASIC, whether running on an IBM PC or connected via a terminal. I realize now that the *Star Trek* game I played on the Wang minicomputer at McGill is closely related to the game *Super Star Trek* in the first *BASIC Computer Games* book; whoever ported it to the Wang had modified it enough that the star map in the game, while still rendered entirely in ASCII graphics, was displayed in the center of the screen rather than scrolling in imitation of a line terminal.

Less aggressive modification might be needed just to get the game working, if your version of BASIC was different from the version the book targeted. For example, many BASICs allowed multiple statements on one line, separated by a colon, which the book took advantage of, but some did not, which would require reworking the code. Or they allowed multiple statements, but separated by a back slash instead.[27] And some differed in their handling of strings, especially the API used to extract subsets of strings. IBM PC BASIC was pretty much in line with what the book expected, except for a minor tweak needed to any code that generated random numbers. This is not too surprising since the standard for the books, which came out in 1978 and 1979, respectively, was Microsoft BASIC—not running on the IBM PC, which did not exist yet, but rather on earlier computers, offering a reminder that Microsoft was in business as a company selling languages for several years before making the deal to sell MS-DOS to IBM that ensured its long-term success. In fact, the reference version for the first book was Microsoft's MITS Altair BASIC, revision 4.0—a descendant of the first product that Microsoft ever sold.[28]

The games were of varying quality. The lack of graphics support left the interactivity a bit lacking. These are the instructions for the *Hockey* game in the first book:[29]

```
QUESTION    RESPONSE
PASS        TYPE IN THE NUMBER OF PASSES YOU WOULD
            LIKE TO MAKE, FROM 0 TO 3.
SHOT        TYPE THE NUMBER CORRESPONDING TO THE SHOT
            YOU WANT TO MAKE.   ENTER:
            1 FOR A SLAPSHOT
            2 FOR A WRISTSHOT
            3 FOR A BACKHAND
            4 FOR A SNAP SHOT
AREA        TYPE IN THE NUMBER CORRESPONDING TO
            THE AREA YOU ARE AIMING AT.   ENTER:
            1 FOR UPPER LEFT HAND CORNER
            2 FOR UPPER RIGHT HAND CORNER
            3 FOR LOWER LEFT HAND CORNER
            4 FOR LOWER RIGHT HAND CORNER
```

And that's how the game went: when one player had the puck, the game would prompt to enter a number of passes (0 to 3), then prompt for the type of shot and area, then randomly generate a result (goal or save), then determine which player had the puck next, and so on until you ran out of fake time (the computer also decided how much fake time each play took).

Believe it or not, these games were compelling to play back in those days (and some of them, to be fair, were better suited to the noninteractive mode), even though you could read the source code and figure out the computer artificial intelligence (AI; from reading the code now, it looks like the slap shot was the best option to score, and the area you aimed at had no effect).

Still, they were examples of large BASIC programs, with *Super Star Trek* from the first book clocking in at over four hundred lines, printed in extra-small type, and *Seabattle*—written by a high school student in Minnesota—from the second book being over six hundred lines.[30] Beyond the unfortunate coding style that BASIC mandated, with subroutines and GOTO targets being line numbers, people continued to use poor variable names; the average number of characters in the book's variable names was perilously close to one.

The author of *Seabattle*, Vincent Erickson, said that the version of BASIC he wrote it for had a limit of two characters for variable names—an obvious impediment to clarity. He wrote the game in 1977, and later entered it in the 1978 Minnesota state programming contest; he then submitted it to *Creative Computing* magazine, which selected it for inclusion in the book, paying Erickson $50 for the rights. To his credit, the program does have a decent selection of comments identifying the different sections of the code, but Erickson didn't remember if he wrote it that way originally or added them when he submitted it for publication. He won second place in the programming contest; on the plus side, he later started communicating via e-mail with a female computer science student at another high school; her teacher had given her a list of entries in the contest. This led to what may be the first marriage between two people who met online.[31]

While anodyne variable names were apparently not enough to block Cupid's arrow, they could be frustrating. There was a checkers game in the first book that contained the logic for the computer AI, which I would have been interested in, but it was buried in a thicket of single-letter variables, and also a maze generation program in the second book, another program whose logic I wanted to decipher—how did it ensure there was only one path through?—yet it was too hard to wade through the code. For the games like *Hockey*, you could use printed strings as a guide; the "somebody scored a goal" logic was clearly in the general vicinity of this:[32]

```
970 PRINT "GOAL " A$(7):H(9)=H(9)+1:GOTO 990
```

But for a game like the *Checkers* or *Maze* programs, those guideposts did not exist. And not surprisingly, comments were almost nonexistent (belated kudos to whoever wrote the *Bowling* program, which at least had a few help-ful comments to delineate different sections of the code).

With all this, it is accurate to say that BASIC, in the form in which it most commonly existed in the early 1980s, was unusable for sharing code between people who did not know each other and did not collabo-rate in detail on the sharing. This is not a knock specifically on the BASIC that Microsoft wrote for the IBM PC, which was quite full featured; it's a fundamental problem with a language that uses line numbers as sub-routine and GOTO targets. John Kemeny and Thomas Kurtz, the inventors of BASIC, recognized these problems, and by the late 1970s had come up with an improved version of BASIC that had properly named subroutines with parameters and other changes that made GOTO mostly unnecessary.[33] Unfortunately the versions of BASIC that shipped with personal computers, which is where most people learned the language, had already split off in their own directions. Kemeny and Kurtz were not happy with the IBM PC BASIC, partly because it reminded them of the earlier incarnations of the language. They described the way it handled numerical variables as "ugly" and "silly," and criticized aspects of the graphics supports as the sign of a "very poorly designed language."[34] Yet there was no closing the barn door after that particular horse had escaped.

At this point in my story I should mention the name Edsger Dijkstra. Dijkstra was a Dutch computer scientist who was born in 1930 (he died in 2002), which positioned him well to invent some of the foundational concepts and algorithms in computer science. Despite looking like a cen-tral casting computer science professor, he had a knack for generating good quotations, and in 1975 wrote a letter titled "How Do We Tell Truths That Might Hurt?" in which he stated, "It is practically impossible to teach good programming to students that have had a prior exposure to BASIC: as potential programmers they are mentally mutilated beyond hope of regeneration." He also labeled Fortran an "infantile disorder" and called it "hopelessly inadequate for whatever computer application you have in mind today: it is now too clumsy, too risky, and too expensive to use."[35] He was silent on the subject of HP calculators, although you should hear what he said about COBOL (another language dating from the 1950s). In any case, BASIC's lack of sophistication made BASIC interpreters simpler, and therefore easier to fit into the minimal memory of early personal comput-ers. Rather than heed Dijkstra's warning, the personal computer industry ordained BASIC as the de facto standard language, contributing to the blight of another generation of programmers. Gates, incidentally, had an almost-identical first experience with computers about a decade before I did, when his school obtained a similar printer-based terminal that connected to a remote computer (except that his first programming language was a dialect of BASIC, not Fortran, so he had to deal only with the mental mutilation, without my hopeless inadequacy added on).

So there I was in fall 1984, a survivor of Fortran and BASIC, heading off to Princeton to major in computer science, preparing to soak up all the software engineering knowledge that had so far eluded me. What happened next, as they say, may surprise you.

2 The Education of a Programmer

Informed by knowledge of what is taught to prospective doctors, lawyers, or accountants, you could not be faulted for picturing my computer science education at Princeton as devoted to instruction on how to design software, reporting on experiments with different languages and methodologies, relaying tips for corralling elusive bugs and hard-to-pinpoint slowdowns, and in general having the faculty impart their combined wisdom on software engineering to the eager assemblage of students.

Before I get into details on why it wasn't quite like that, I want to cover a bit of terminology. People who write software programs are called *programmers*. They are also called *developers* or *software developers* as well as *software engineers*, *software development engineers*, or sometimes *software design engineers*. I called myself a programmer when I was younger, while at Microsoft we were informally called developers (often shortened to "devs"), but my title was software development engineer. One of the questions in this book is whether the word *engineer* belongs there. But for now, consider them all to be interchangeable.

Meanwhile, people who go to college to study programming usually major in *computer science*—another two-word phrase that may not yet have earned its second word—but they may instead major in *software engineering*. Some claim there is a distinction between those two, with computer science more focused on theory and software engineering more concerned with the application of that theory, yet there is no agreement on the difference, or whether it exists at all, so treat those two as equivalent too.

In any case, off I went in 1984, home-brewed programming experience in hand, to Princeton University to study computer science. Princeton's was a typical high-level computer science program: good facilities, smart students, and professors who were recognized as leading authorities in their areas of research. But the professors' areas of research were primarily related to theoretical computer science, essentially the study of algorithms (Princeton had a reputation for being a bit more algorithm-focused than other schools,[1] although in my observation of the graduates of various other schools, when I was later interviewing them for jobs at Microsoft, I did not notice any difference in the training they had received). There was only one class that centered on how to write software: the introductory computer science class that I took my first year, where we were taught a language called Pascal.

The big advance that Pascal had over early 1980s' BASIC was that it supported passing parameters to its version of subroutines, which it called *procedures*. (Pascal made a somewhat-unnecessary distinction between procedures, which did not return a value, and *functions*, which did. I'll use *procedures* here; both would fall under the more language-independent term API.) In addition, any variables declared inside a procedure were *local variables*, which means you didn't need to worry about whether they had the same name as a variable declared outside the procedures. This made it possible to call a procedure written by somebody else without knowing the details of how it was implemented—the basis for building up code in layers. This is essentially impossible in IBM PC BASIC with its line-number-based subroutines with no parameters, and all variables being global.

My sister, who is a year older than me, also took a Pascal course in college. I once got into a debate with her about whether it was important to be able to define procedures with parameters, as Pascal allowed, or whether BASIC's support for unnamed, unparameterized subroutines was sufficient. This latter position is hopelessly naive in retrospect, but nonetheless it was the side I chose. At this point my sister's programming experience in Pascal consisted of the typical short assignments (sort an array of integers, etc.) that I had cut my teeth on when learning WATFIV, whereas I had written several games in BASIC of decent complexity (with what would be described today as 8-bit graphics, although since the IBM PC only supported 16 colors, they actually were 4-bit graphics). There is no doubt in retrospect that I was wrong and my sister was right. With sufficient care you could work around the problems of not having named subroutines and local variables, but they are such a convenience and avoid so many preventable mistakes that dismissing them is indefensible. In my defense, recall Dijkstra's comment that BASIC programmers are "mentally mutilated beyond hope of regeneration."[2] Perhaps at that point my brain was so torn up inside from my intense exposure to BASIC that I couldn't think straight and lost an argument to my sister.

In my introductory class at Princeton, we learned the basics of Pascal and wrote the simple programs used to learn a language. Our textbook, *Programming in Pascal* by Peter Grogono,[3] did a thorough job of explaining the syntax of Pascal, without spending much time on what you should do with it, and I don't recall the instructor getting into too many details either. As Harlan Mills, a noted writer on software topics who spent over twenty years managing teams at IBM, once wrote,

Our present programming courses are patterned along those of a "course in French Dictionary." In such a course we study the dictionary and learn what the meanings of French words are in English (that corresponds to learning what PL/I or Fortran statements do to data). At the completion of such a course in French dictionary we then invite and exhort the graduates to go forth and write French poetry. Of course, the result is that some people can write French poetry and some not, but the skills critical to writing poetry were not learned in the course they just took in French dictionary.[4]

The trend in programming circles at that time was *structured programming*. Donald Knuth is a longtime professor of computer science best known for the multivolume opus *The Art of Computer Programming*, a comprehensive summary of software algorithms that he began working on in 1962. He wrote, "During the 1970s I was coerced like everybody else into adopting the ideas of structured programming, because I couldn't bear to be found guilty of writing *unstructured* programs."[5] It's likely that for any language then in use, there existed a book whose title contained the word *structured* followed by the name of the language. You could find *Structured Programming Using PL/1 and SP/k* (1975), *Structured Programming in APL* (1976), *Programming in FORTRAN: Structured Programming with FORTRAN IV and FORTRAN 77* (1980), *Structured COBOL: A Pragmatic Approach* (1981), *Problem Solving and Structured Programming in Pascal* (1981), *Structured Basic* (1983), and so on.

Gerald Weinberg is another longtime observer of the software landscape who, among other things, worked for IBM on the software for Project Mercury, the US program to put an astronaut into space in the early 1960s. In his foreword to the 1976 book *Structured Programming in APL* (APL was another programming language, whose name is unrelated to the term API), Weinberg lays out the structured manifesto with his usual flourish:

APL has earned such a reputation for disorderly conduct that "structured APL" rings as off-key as "immaculate pigsty" or "honest politician." Yet we must not blame the language for the disorderly conduct of its users—or misusers. In the hands of responsible and properly educated programmers, APL becomes a marvelously disciplined tool, a tool unlike any other programming language in common use.

The problem, of course, lies in the phrase "properly educated." For too long, in too many places, APL users have learned the language "in the streets," as any examination of their programs would show. Their textbooks are little more than reference manuals, and offer no corrective to the worst effects of the oral tradition.[6]

Learning "in the streets," textbooks as "little more than reference manuals"—indeed! Weinberg was making the same point I am making in this book, forty-plus years later: most programmers are not properly educated in how to program, and it shows in their code.

It was unclear whether structured programming was a process—a structured approach to producing a program—or result—a program that is structured, no matter how it got that way. Knuth's and Weinberg's quotes above make it sound like it is the second. I concur; in the end, the code is what remains as well as what will determine how quickly a new programmer can figure out how a program works. Nonetheless, the literature, while clearly enamored of the term, varied a lot in deciding what structured programming really was.

Structured COBOL: A Pragmatic Approach gets to page ninety-five before providing a brief section on structured programming, explaining, "This is the first mention of the term *structured programming*, although every program presented so far has been 'structured.' Structured programming is the discipline of making a program's logic easy to follow. This is accomplished by limiting a program to three basic logic structures: sequence, selection, and iteration."[7] The book then presents flowcharts for each of those three logic structures. *Sequence* here just means "one program statement following another"; *selection* means IF statements and the resulting choice from those; and *iteration* is loops in all their various forms.

Structured BASIC devotes one six-page chapter to structured programming, about halfway through the book, which starts out by stating, "Structured programming is an additional approach to program development that usually results in more efficient code, less time spent on development, program logic that is easier to follow, and a resultant program that is easier to debug and modify."[8] It's hard to argue against that, but the approach that the authors present is a mix of "structure charts," which are a visual representation of the different parts of a program, plus flowcharts that indicate the same three basic concepts in programs (sequence, selection, and iteration), so clearly the authors view structured programming as "a structured approach to producing programs." They do briefly mention, at the end of the chapter, that comments can be helpful, and that indenting IF and FOR blocks can help with readability—the only nod toward structuring of the actual code.[9]

Meanwhile, *Structured Programming in APL*, despite Weinberg's rousing introduction, waits until the epilogue before devoting two and a half pages to the topic of structured programming (to be fair, the book does use structure diagrams extensively), starting out with this: "Perhaps, in closing, we should mention something about the mysterious phrase 'structured programming,' which appears in the title, but nowhere else. At the time the book is being written there is still some controversy about exactly what structured programming is. But there is no disagreement about the fact that it is valuable."[10] What follows is a hand-wavy definition that encompasses the differences between engineering a bridge and engineering software, the fact that software is often modified from its original purpose, structure charts and the sequence-selection-iteration trinity, the importance of design, and the right length for a program; it also includes the sentence "there is still controversy over whether to use names (for variables and labels and programs) that are meaningful or meaningless."[11]

This is all fine, but it's incredibly basic: *all* programs in high-level languages, however structured they claim to be, consist of sequences of instructions, selection by IF statements (or their equivalent), and iteration in loops. At the bottom level, those are the bricks from which software is built. If this is structured programming, it's hard to imagine what territory is left for unstructured programming to claim.

Structured Programming Using PL/1 and SP/k probably provides the best summary:

Certain phrases get to be popular at certain times; they are fashionable. The phrase, "structured programming" is one that has become fashionable recently [the book came out in 1975]. It is used to describe both a number of techniques for writing programs as well as a more general methodology. ... The goals of structured programming are, first, to get the job done. This deals with *how* to get the job done and how to get it done *correctly*. The second goal is concerned with having it done so that other people can see how it is done, both for their education and in case those other people later have to make changes in the original program.[12]

The authors also offer a diplomatic warning about GOTO statements (which are written as GO TO in PL/I): "Since computer scientists came to recognize the importance of proper structuring in a program, the freedom offered by the GO TO statement has been recognized as not in keeping with the idea of structures in control flow. For this reason we will *never* use it."[13]

On a related note, at the end of *Structured COBOL*'s discussion of structured programming, it offers the following:

Conspicuous by its absence in Figure 6.1 [which lays out flowcharts for sequence, selection, and iteration] is the GO TO statement. This is *not* to say that structured programming is synonymous with 'GO TO less' programming, nor is the goal of structured programming merely the removal of all GO TO statements. The discipline aims at making programs understandable, which in turn mandates the elimination of indiscriminate page turning brought on by abundant use of GO TO. ... [U]nstructured programs often consist of 10% GO TO statements.[14]

Now we are getting somewhere, and I think the authors doth protest too much: when you boil down the difference between structured and unstructured programming, what remains is getting rid of GOTO.

What is so bad about GOTO?

The ever-voluble Dijkstra wrote a letter to *Communications of the ACM* in 1968 titled "Go To Statement Considered Harmful." The letter's title sounds suitably Dijkstra-esque, but he later claimed that it was provided by Niklaus Wirth, the inventor of Pascal, who was the editor of the magazine at the time; Dijkstra's original title was the less provocative: "A Case against the Go To Statement."[15] The letter begins,

For a number of years I have been familiar with the observation that the quality of programmers is a decreasing function of the density of **go to** statements in the programs they produce. More recently I discovered why the use of the **go to** statement has such disastrous effects, and I became convinced that the **go to** statement should be abolished from all "higher level" programming languages (i.e., everything except, perhaps, plain machine code).[16]

Dijkstra's insight is that a source code listing is static, but what we care about, when figuring out what a program does and if it is correct, is the state of the computer while executing it (what he calls the "process"), which is dynamic. He points out that

our intellectual powers are rather geared to master static relations and that our powers to visualize processes evolving in time are relatively poorly developed. For that reason we should do (as wise programmers aware of our limitations) our utmost to shorten the conceptual gap between the static program and the dynamic process, to make the correspondence between the program (spread out in text space) and the process (spread out in time) as trivial as possible.[17]

In other words, while reading the code it should be as easy as possible to keep track in your mind of the state of all the variables and what they mean as the computer is executing a given line of code. Dijkstra then explains that with sequencing, selection, and iteration, it is relatively easy to figure out the state of the process at any point in the code, but when you allow the code to arbitrarily jump to any other location via a GOTO, it is hard to know the state of the process as it passes through the targeted location because that location now has multiple ways that it can be reached, and you can't know what state the process was in at all the different points from which it may make that jump. Dijkstra concludes, "The **go to** statement as it stands is just too primitive; it is too much an invitation to make a mess of one's program."[18]

Mills makes a similar observation: "In block-structured programming languages, such as Algol or PL/I, such structured programs can be GO TO–free and can be read sequentially without mentally jumping from point to point." (He continues, though, by repeating the official story that "in a deeper sense the GO TO–free property is superficial. Structured programs should be characterized not simply by the absence of GO TO's, but by the presence of structure.")[19]

The argument against GOTO was bolstered by an academic paper laying out the Böhm-Jacopini theorem, which proved that any program could be written without GOTO statements.[20] The proof relied on a somewhat-contorted programming style; in particular, you wound up using extra variables to avoid certain GOTO statements. While inside a loop, it is frequently useful to exit the loop early, before you have finished every planned iteration. An example would be code to look for something in an array, here presented in the language C# (pronounced "C sharp"). The first line is equivalent to the FOR J = 1 TO 10 loop that we saw in our BASIC illustration in the last chapter, but rewritten in C# (and looping from 0 to 9 instead of 1 to 10):

```
for (j = 0; j < 10; j++) {
    // is the j'th element of the array
    // the one we want?
```

```
    if (this_is_the_one(j)) {
        // if it is, then we can exit the loop
        goto endloop;
    }
}
endloop:
```

The GOTO statement jumps to the named label (endloop), avoiding unnecessary iterations through the loop, and at the end the value of j tells you which element of the array you want. Without this GOTO, per the formal Böhm-Jacopini theorem, you need to add an extra variable to short-circuit the unneeded array iterations, and another one to keep track of where it was found:

```
foundit = false;
foundlocation = 0;
for (j = 0; j < 10; j++) {
    if (!foundit) {
        if (this_is_the_one(j)) {
            foundit = true;
            foundlocation = j;
        }
    }
}
```

This is cheesy and harder to read than the first version with the GOTO statement. In fact, many languages (including C#, for those of you grinding your teeth at the code above) have a statement called BREAK that executes "exit the loop now" without needing the label (and the shunned GOTO statement), making this much cleaner:

```
for (j = 0; j < 10; j++) {
    // is the j'th element of the array
    // the one we want?
    if (this_is_the_one(j)) {
        // if it is, then we can exit the loop
        break;
    }
}
```

Nonetheless, some languages consider the BREAK statement (and related CONTINUE statement, which skips only the remainder of the current iteration of the loop, not all future iterations) to be a GOTO in disguise and don't allow it. Pascal, back in those days, was one of those purist languages, with Böhm-Jacopini available as justification; *Programming in Pascal* has an instance of skipping unneeded loop iterations using an extra variable (which it calls a *state variable*), and the same example using a GOTO, and decides that the state variable is better: "Although this example illustrates the effect of the goto statement, it does not justify the use of the goto statement."[21]

Personally I find code with a BREAK in it (or even a GOTO to a clearly labeled "loop end" label) much easier to read than code that adds an extra variable to avoid using it. The problem is not so much these "nearby" GOTOs but rather the more indiscriminate use in which the program jumps all over the place, such as you see in the *BASIC Computer Games* books or DONKEY.BAS. (As an extra bonus, in BASIC, if you had a subroutine starting at, say, line 700, but no GOTO at line 690 that skipped past the subroutine, the BASIC interpreter would roll right into line 700 and start executing your subroutine, with whatever state the relevant variables happened to be in; one of the benefits of Pascal and other languages that formally declare procedures is that they avoid this problem, since the procedure code is not part of the main code path.)[22] This GOTO-laden style of programming was derided as "spaghetti code" because trying to follow the path taken through the code was like trying to follow a single strand of spaghetti in a bowl; it would disappear into unknown places and then reappear somewhere else, with no clarity on what exactly happened in between or even whether you were following the same strand. Kemeny and Kurtz, the inventors of BASIC, acknowledged that the effect of every line of code having a line number, which therefore made every line of code available as the potential target of a GOTO, was the "one very serious mistake" they made in the design of the language.[23]

If GOTO statements are so terrible, you might wonder why people used them in languages where they were not necessary. The authors of *Fortran with Style* explained in 1978: "The unconditional transfer of control, which is the function of the GOTO statement, has been associated with programming since its inception. Its historical ties have left indelible marks on today's programming languages."[24] While high-level languages are built up from selection and iteration (IFs and loops), assembly language has lower-level building blocks, which you may recall from the last chapter: moving data between registers and memory, performing operations on registers, comparing registers, and jumping to other locations in the program. That "jumping to other locations in the program" is a GOTO (although the term *jump* is usually used in assembly language), and a higher-level-language construct like an IF is built up using jumps in assembly language; when reading code in a higher-level language, your eye will automatically slide down past the block of code that follows an IF test, but in assembly language you have to explicitly jump past it.

As Mills explains in his 1969 essay "The Case against GO TO Statements in PL/I" (whose opinion on GOTO can be accurately inferred from the title), for a programmer coming from assembly languages, jumps are a natural thing, and you can't write any reasonable program without them.[25] It makes sense that assembly-language programmers, when moving to a higher-level language, would not differentiate jumps done in the context of selection and iteration, which arguably are still "structured," from jumps to random places in the program. Mills exhorts his audience, "It might not be obvious ... that GO TO's could be eliminated in everyday PL/I programming without its being excessively awkward or redundant. But some experience and trying soon uncover the fact that it is quite easy to do; in fact, the most difficult thing is to simply decide to do it in the first place."[26]

He also states, "It is not possible to program in a sensible way without GO TO's in FORTRAN or COBOL. But it is possible in ALGOL or PL/I."[27] Our "sum up the numbers" Fortran program from chapter 1 had two GO TO statements for a very simple algorithm. In the more modern language Pascal, this could be written without them as

```
var sum, x: integer;
begin
    sum := 0;
    repeat
        read(x);
        sum := sum + x
    until x = 0;
    writeln(sum);
end.
```

What about the books that claimed to teach structured programming in Fortran and COBOL?

Programming in FORTRAN: Structured Programming with FORTRAN IV and FORTRAN 77 describes the well-known troika, with some names changed: "Three basic control structures are sequence (begin-end), decision (if-then-else) and loop (while-do). These, sufficient to present any algorithm, constitute the fundamental means of a systematic programming process called *structured programming*."[28] But wait! Those are the theoretical constructs; Fortran doesn't actually have a while-do loop, so the book explains how to write one using GOTO statements.[29] The begin-else control structure merely involves putting statements one after the other, so that subject is never broached again. For if-then-else, later versions of Fortran do provide reasonable support, whereby you can have a set of lines of code that runs when the IF condition is true, and another set that runs when it is false, which is known as supporting *block IFs*. But earlier versions only let you run one statement when the IF condition was true, so that statement was perforce often a GOTO (as our "sum up the numbers" Fortran did). The book points out, "The old FORTRAN standard did not contain the block IF construct; thus it is not available in FORTRAN IV or in such compilers as WATFOR and WATFIV."[30] Yes, I must confess that the first programming book I ever read was for a language variant so antediluvian that it didn't even have block IF statements.

Meanwhile, *Structured COBOL: A Pragmatic Approach* has it slightly easier. COBOL does have block IF statements, and a loop construct called PER-FORM UNTIL, although it requires you to put the body of the loop into a separate procedure that makes it harder to read,[31] and COBOL suffers from so much other clunkiness that I can see why Mills threw it under the "not structured" bus; you can't program in a "sensible way" in COBOL no matter what parts of the language you use. Nevertheless, the authors can get away without using GOTO statements, except in one specific case: when a program wants to exit, they use a GOTO to jump to the end of the program. As they write about one program listing, "Figure 11.2 also contains five 'villainous' GO TO statements, but their use is completely acceptable (to us, if not to the most rigid advocate of structured programming)." They further explain, "The authors maintain that a structured program can include *limited* use of the GO TO, provided it is a forward branch to an EXIT paragraph" (this must be the pragmatism of the book's subtitle manifesting itself).[32] I agree;

if you are reading through the program and get to the point where it is about to exit, there is no mental model that needs to be maintained. My copy of this book used to belong to my mother, from a programming class she took in 1983. In the margins of the book, next to the first quote, she wrote, "Hear hear," and next to the second, "God will forgive you!" Knuth wrote in favor of allowing GOTO for "error exits," and even Dijkstra, in his anti-GOTO screed, conceded that "abortion clauses" or "alarm exits," meaning this same sort of "jump to the end of a block of code" approach, might be acceptable.[33]

Anyway, let's say that structured programming just means "use GOTO as little as possible." Clearly, from my hopeless argument with my sister, even this lesson, which in hindsight seems obvious, needed to be drilled into habitués of Fortran, COBOL, or BASIC. I suppose I could say that I learned structured programming at Princeton in that I did absorb the lesson about eschewing GOTO. But that's probably one of the few lessons I was explicitly taught there. Because in high school I had succeeded in teaching myself programming and accomplishing reasonable results with it, I was extremely confident that the way I had learned things was correct, despite having no real basis for this claim beyond my own experience.

The logician Raymond Smullyan proposes in his book *What Is the Name of This Book?* that people are either conceited or inconsistent:

A human brain is but a finite machine, therefore there are only finitely many propositions which you believe. Let us label these propositions *p1*, *p2*, ..., *pn*, where *n* is the number of propositions you believe. So you believe each of these propositions *p1*, *p2*, ..., *pn*. Yet, unless you are conceited, you know that you sometimes make mistakes, hence not everything you believe is true. Therefore, if you are not conceited, you know that at least one of the propositions *p1*, *p2*, ..., *pn* is false. Yet you believe each of the propositions *p1*, *p2*, ..., *pn*. This is straight inconsistency.[34]

Smullyan's point was that a reasonably modest person is behaving inconsistently, which he happily admits to; when it comes to programmers, however, the conceited approach usually wins out.

Once my introductory Pascal class was over, and I had learned to appreciate the value of passing parameters to named procedures, the rest of the undergraduate classes that I took dealt with more specific topics: how to design a compiler, how a virtual memory manager worked, and how three-dimensional graphics were projected onto a two-dimensional display—all interesting, but those classes all focused on the specific algorithms needed for those problems, and since I have never worked on those areas in my professional career, it's not knowledge that I draw on in my everyday work. Nobody taught us how to design large programs and get them working on a deadline. We were given assignments that required large programs along with a deadline to get them working, and we made it happen as best we could.

Sophomore year was the first time I took a class where I used the programming language called C, which I wound up using for most of my college and professional career. After the professor explained the goals of the first assignment, a student rather hesitantly raised their hand and asked how we were supposed to learn C? No problem, said the professor, use this book (it was the original *The C Programming Language*, written by the language's inventors, Brian Kernighan and Dennis Ritchie). I learned C by reading the book, looking at examples to try to discern their underlying motivation, and most important, trying things and fixing them when they didn't work—the same process I had used to learn IBM PC BASIC four years earlier.

As Mills put it, the book taught me the dictionary. The rest of it, the bulk of what I learned about software engineering—how to split a big problem into smaller ones, how to connect the pieces together, how to figure out why it didn't work, and how to decide when it was finished—I figured out on my own by trial and error, and everybody else in the class figured it out on their own with their own trials and errors.

And as a final point, I had only one class that involved modifying a program that somebody else had written; the rest of my projects were all greenfield ones, in which I started from scratch. Modifying existing code is what a professional programmer spends the vast majority of their time doing, but my work at school gave me little preparation for sitting down with a large program and figuring out what the heck the original author was thinking—or if they had done something right, using their code as an example.

I have some code saved from my time at Princeton (what, you say you don't have thirty-year-old printouts stashed in your attic?), which I can look at now through the lens of the intervening time spent earning a living as a programmer. It's what I would expect—a projection of my BASIC experience onto C (except without GOTOs): short variable names that don't clarify their meaning, no comments to explain what is going on or delineate different areas of the program, and repeated code that should have been pulled into a shared function (which is the term C uses for an API). I assume the code worked, although I would be hard put to verify that now by reading it. It served its purpose: to procure a grade for a class assignment and then never be looked at again.

How are all these stories about my education related? The common theme is that in all cases, I was *self-taught*. In high school I was fairly evidently learning on my own. But even my Princeton years are deceptive. A casual observer would note that I was taking a lot of computer science classes and I was learning a lot about how to write software. The second was a by-product of the first, though, and not a direct result. What was missing was anybody explaining what I should do *before* I did it wrong a couple of times, or anybody looking over the details of how I had written a program as opposed to the result that it achieved. Despite graduating with a degree in computer science, I was sorely lacking in the wisdom that I would eventually acquire, through experience, during my career as a programmer.

And it's not just me: essentially all programmers working today were self-taught. The people who designed the Internet were self-taught, those who architected Windows were self-taught, and the people who wrote the software that is running on your microwave oven were self-taught too.

What does this mean for software engineering? The most obvious issue is that it is incredibly wasteful to have everybody figure things out from square one (or perhaps squares two and three), over and over and over again. The notion of experimenting and using the results to inform further steps, building up an engineering process on the work of those who have gone before, is almost completely absent from software development. We're not standing on the shoulders of giants; at best they are offering us a knee up as a boost. The instructor Scott Bain, who teaches at a company called NetObjectives that offered training at Microsoft, once pointed out that there is no well-defined path to becoming a software engineer: you don't go to college and major in a certain subject, then take a set of well-determined certification tests, then do an apprenticeship, and then become certified. You can do the college major thing, but once that's done you hang up your shingle and say you are a programmer, and hope a company fishes you out of the ocean of similar people. And worse yet, it may be that your first year in college is already too late to start down the path, if you haven't spent the last couple years in high school hacking away in your basement.

That makes it hard for people who want to hire programmers (either to work at a software shop like Microsoft or as consultants on a project for a business) to figure out who is qualified. But the subtler effect is that it can scare off people who are considering becoming programmers. How do you embark on the path if it isn't well defined? Do you have to devote yourself at a young age to poring over programming manuals in your spare time? If you *weren't* a member of the programming club in high school, are you permanently behind?

And the largest, most important group that it scares off is women.

In 2002, a fascinating book appeared: *Unlocking the Clubhouse* by Jane Margolis and Allan Fisher. The book studied students in Carnegie Mellon University's highly respected computer science program as a basis for understanding why women are underrepresented in the industry. Although the female computer science students all appeared highly qualified, and arrived at college motivated and confident, many of them soon experienced a similar sense of inferiority. Here is a sample of quotes from students:

Then I got here and just felt so incredibly overwhelmed by the other people in the program (mostly guys, yes) that I began to lose interest in coding because really, whenever I sat down to program there would be tons of people around going, "My God, this is so easy. Why have you been working on it for two days, when I finished it in five hours?"

I'm actually kind of discouraged now. Like I said before, there are so many other people who know so much more than me, and they're not even in computer science. I was talking to this one kid, and ... oh my God! He knew more than I do. It was so ... humiliating kind of, you know?

What am I doing here? So many other people know so much more than me, and this just comes so easy to some people. ... It's just like there are so many people that are so good at this, without even trying. Why am I here? ... You know, someone who doesn't really know what she is doing?[35]

I don't think the men playing the role of "other people" in these quotes had an innate ability to write software; it's that they had been practicing much longer than the women (which doesn't excuse them for making fun of someone for taking more time to finish a program). I've discussed this topic with female computer science students and heard similar comments— in fact, almost stunningly similar: the same discouraging sense that the people who had been hacking away in high school knew so much more, and were so much more capable and prepared for future success (and as a former high school hacker who majored in computer science, I was inadvertently complicit in creating the equivalent environment at Princeton). The sameness of the quotes might suggest a glimmer of hope that at least these women could find solace and support in each other, and wage a determined battle against the propeller-heads. But it appears that the mental anguish was a lonely, internal battle, with each individual mind beset by nagging questions that caused these intelligent, motivated, capable women to repeatedly doubt their abilities, until one by one they dropped the fight and majored in another subject.

The underlying problem begins at an early age, according to Margolis and Fisher: "Very early in life, computing is claimed as male territory. At each step from early childhood through college, computing is … actively claimed as 'guy stuff' by boys and men. … The claiming is largely the work of a culture and society that links interest and success with computers to boys and men." They write,

Despite the rapid changes in technology and some fifteen years of literature covering the era of the ubiquitous personal computer, a remarkably consistent picture emerges: more boys than girls experience an early passionate attachment to computers, whereas for most girls attachment is muted and is "one interest among many." … Developing and exploring the computer are truly epiphanies for many of these male students. They start programming early. They develop a sense of familiarity; they tinker on the outside and on the inside, and they develop a sense of mastery over the machine.[36]

In addition, "Girls at nine and ten are feisty, filled with spirit and confidence, but as puberty hits, they begin to pull within themselves, doubt themselves, swallow their own voices, and doubt their own thoughts."[37] The problem is exacerbated in the years leading up to college: "In secondary schools across the nation, a repeated pattern plays out: a further increase in boys' confidence, status and expertise in computing and a decline in the interest and confidence of girls. Curriculum, computer games, adolescent culture, friendship patterns, peer relations, and identity questions such as 'who am I?' and 'what am I good at?' compound this issue."[38]

Computer science is not the only field in which women may receive societal messages that steer them away during high school. And certainly there are areas, particularly sports, where success as a professional almost always requires dedicated commitment and interest in high school, if not earlier. Yet computer science packs a one-two punch because, currently, it can be self-taught at a young age: women are losing interest in the field at precisely the same time that men are not only cultivating their interest but also learning the actual skills that propel them to a successful career, which makes it much harder to catch up in college.

In my 1988 graduating class at Princeton, according to the alumni directory, five out of forty-one computer science majors were women.[39] I know one of the women had not programmed before arriving on campus, although most if not all of the others had experience similar to mine. It's a small sample size, but more important, those are the five who stuck it out until the end; I don't know how many other women started down the path to major in computer science but then changed their minds after the types of dispiriting experiences described in *Unlocking the Clubhouse*. There may also have been male computer science majors or former computer science majors who were new to programming when they arrived at the school, although every male whom I can recall discussing the topic with had been writing programs in high school.

During the time I was at Princeton, there was a campus computer network, and the school attempted an early experiment at extending it into dorm rooms. The only problem was that a year elapsed between when the administrators asked who was interested and when they ran the network cables, so you only had a network connection if the person who had lived in your room the year before had requested one. Even then, what was available on the network was quite limited; there were no websites (the World Wide Web and associated protocols not having been invented yet), so you could only connect to a few mainframe computers in an updated version of "Adam in his parents' bedroom with the line terminal." And even the people who had computers in their rooms had PCs, which were different from the computers on which we could work on our assignments. So for this variety of reasons, all programming for my classes was done in a computer lab in a building named after John von Neumann, the famous mathematician who had worked at the Institute for Advanced Study near the Princeton campus.

"The Neum," as we called it (rhymes with … nothing much, but the vowel sound is the same as the word "boy"), was a below-ground bunker, whose roof, according to legend, had a half-inch-thick slab of iron embedded in it to prevent enemy powers from spying on the computers inside. I'd crank out my programming assignments during nighttime coding jags, fortified by a gallon of Wawa iced tea and a foot-long bacon cheesesteak from Hoagie Haven (which closed at midnight, so you had to plan ahead to lay in your provisions). Inside were long tables with computers (video terminals, actually)—a precursor to the open workspace that many software companies use today on the pretext of the better sharing of information, but in this case it was just the easiest way to arrange them.

Working elbow to elbow with my fellow grunts should have at least given us an opportunity to learn from each other, and possibly for the few women in the class to offer support to each other, but I don't remember this happening; we mostly coded away in grim, solitary silence (one classmate had a girlfriend who would sit quietly next to him while he worked, which must have been somewhat boring for her and slightly nerve-racking for him). Even when I worked on a project with a partner, we usually split the work up and tackled it independently. So the benefit I got from Princeton was not the learning from my peers that was so helpful in other classes. It was that I was forced to *write* a lot of programs, giving me ample opportunity to figure out how to write them, debug problems, and fix those, but all on my own. I had a job working at the computer center, where we would staff various locations to answer questions, yet it was well known that working at von Neumann was an easy shift because nobody ever asked any questions. Many of us had been self-taught in high school and continued to self-teach ourselves in college.

This is ridiculous, right? The world depends on software, but are software skills really gained in marathon programming sessions as a teenager? In trying to imagine a similar situation in the field of medicine, I picture a student at a medieval medical school writing a letter home, complaining, "Everybody else is so much better at using leeches than I am ... and this one kid! Back home he's already performed three pocketknife amputations!" Indeed, it was common in the United States two hundred years ago for doctors to learn their trade by apprenticing themselves to an established doctor, but eventually the need for formal education was recognized.[40] The fact that today's programmers can on their own acquire such a head start in knowledge—not useless knowledge, but the same knowledge that was being learned (by a process independent from formal instruction) in school—is an indictment of the discipline of software engineering, not of the inexperienced students.

Several of my classmates at Princeton wound up working at Microsoft too; I once had a discussion with one of them, who had also written software in his spare time in high school, about how we really should have skipped Princeton and gone to work for Microsoft directly after high school. We were kidding, but there was a large nugget of truth in there. Assuming Microsoft would have hired us back then, we would have emerged in 1988 with a lot more experience (and money) than we did after college, and most important, we weren't much less qualified in 1984 than we were in 1988. This is partly related to a onetime historical window, because in 1984 there weren't a lot of people out there who had several years of experience programming on the IBM PC. But the primary reason is that as preparation for developing large pieces of commercial software, writing video games in BASIC was almost as useful as going to college and majoring in computer science. And certainly by 1988, four years of working at Microsoft would have left us far more qualified than if we'd spent the last four years earning computer science degrees. While it would be unthinkable today for a doctor, say, to skip medical school and go straight to practicing medicine, no such gap exists for programmers. Microsoft would occasionally hire a developer who had majored in something like music and watch them be as successful as the computer science majors, which was impressive for them personally, but a little strange if you think about it.

Yet this is roughly where we stand with software education today. In 2011, George Washington University professor David Alan Grier wrote in *IEEE Computer* magazine, "It isn't necessary to have a bachelor of science degree to be considered a software engineer. According to the Bureau of Labor and Statistics, a software engineer is the leader of a programming or system development project, not necessarily a trained engineer. ... Only in some cases, notably the most restrictive professions, does it consider 'skills, education and/or training needed to perform the work at a competent level,'" with software engineer apparently not making the restrictive professions list. Grier then concludes, "Those who can do the work, no matter how they may have been trained, can generally find work."[41] People *can* gain a big leg up on their college computer science education by learning in their spare time, on their own. And unfortunately, this knowledge is often acquired at a stage of people's lives where women, for whatever reason, tend to be less into computers than men. And excluding half the planet from your talent pipeline certainly affects your ability to hire all the qualified people you want.

All this raises the question, Why has the software industry continued to operate this way?

The software industry has evolved in just a couple of generations, leaving little time to reflect on how things are done. As Mills again notes, from 1976,

In the past twenty-five years a whole new data processing industry has exploded into a critical role in business and government. Every enterprise or agency in the nation of any size, without exception, now depends on data processing software and hardware in an indispensable way. In a single human generation, several hardware generations have emerged, each with remarkable improvements in function, size, and speed. But there are significant growing pains in the software which connects this marvelous hardware with the data processing operations of business and government.

Had this hardware development been spaced out over 125 years, rather than just 25 years, a different history would have resulted. For example, just imagine the opportunity for orderly industrial development with five human generations of university curriculum development, education, feedback for the expansion of useful methodologies and pruning of less useful topics, etc. As it is, we see a major industry with minimal technical roots.[42]

More important, however the software sausage is made, it is so incredibly useful that there has not been much pressure to improve how things are done. In an environment where customers are clamoring, "Give me more of that sweet, sweet software!" the industry has no real incentive to step back and try to improve things.

Fundamentally, people in the software industry see nothing wrong with being self-taught because, hey, it worked for them. Weinberg once wrote,

Another essential personality factor in programming is at least a small dose of *humility*. Without humility, a programmer is foredoomed to the classic pattern of Greek drama: success leading to overconfidence (*hubris*) leading to blind self-destruction. Sophocles himself could not have invented a better plot (to reveal the inadequacy of our powers) than that of the programmer learning a few simple techniques, feeling that he is an expert, and then being crushed by the irresistible power of the computer."[43]

Unfortunately, humility is not something that programmers tend to have thrust on them. Which brings us to the real problem with programmers being self-taught: it makes them arrogant.

And why not? By dint of sheer brainpower, without ever having to go through an apprenticeship, pay their dues, submit to any standardized certification, or even get a relevant college degree, programmers have arrived at a situation where they can be paid large sums of money to pursue an activity that many of them would do in their spare time anyway, in an environment that entails no undue physical exertion or risk. What better validation could there be of their own greatness?

We'll keep this thought in mind as we dig into what it's like to work as a professional programmer.

3   Layers

I didn't become a programmer until a year after college.

By that point I had been writing programs for a decade, majored in computer science, and worked at a small software start-up for a year. Unbeknownst to me, that was just practice for my final test.

The fateful sequence of events began when my manager called me into his office. My company, Dendrite Americas, was writing software that allowed representatives of pharmaceutical companies, armed with laptops, to plan their sales meetings with doctors. Every night they would dial in to our central computer to upload notes they had gathered during the day and then download updates to their database of doctors—quite advanced for the late 1980s. In certain cases, in no discernible pattern, the street address of one doctor was being replaced with that of a different doctor. Nobody else had been able to figure out what was going on, and my manager wanted me to give it a try.

Being selected for this assignment, from among the fifteen or so programmers at the company, was a compliment. The sheriff in an old Western movie was choosing his posse and telling me, "Tex, you're the best shot we've got." Nonetheless, I felt my stomach sink—a feeling that I had never felt when facing a programming task. The challenging part in situations like this is not so much fixing the problem as it is finding it, and I was nervous about whether I would be able to find it.

Bugs in software are described by *repro steps*: the sequence that the user follows to reproduce the bug, as in, "Run spell-check, then try to save the document, and you'll get an error." They can be broadly divided into two categories: bugs that happen every time, and bugs that happen only sometimes, despite following the same repro steps. Those that happen every time are vastly preferable, at least from the point of view of a programmer trying to solve them, because if you can make the bug happen reliably, you can eventually narrow down where the problem is. Like the annoying rattle in your car that goes away when the mechanic is listening, software bugs that occur intermittently make you pull your hair out. In reality, even intermittent bugs do happen "every time"; it's just that a certain set of factors have to come together, and the repro steps, as best as they are known, are not detailed enough to always trigger the exact situation.

There is another way to divide software bugs: bugs in a program that you wrote yourself, and bugs in somebody else's program. When the bug is in somebody else's program, you know nothing about the details, so you are starting at square one, or line one. The problem I was being tapped to fix was in the worst quadrant: an intermittent bug in somebody else's program. This was a new experience for me. Previously, in high school and college, I had rarely worked with code written by somebody else. And even when I had, the data I was working with was small enough, and the programs I was working with were simple enough, that any bug was easy to reproduce.

Throw in the pressure of being put on the spot, with paying customers waiting for a fix while their salespeople wandered aimlessly around New Jersey, and this was the moment when I was going to earn my stripes as a programmer.

If you watch home improvement shows, you have no doubt seen the knob-and-tube reveal, in which the contractor informs the homeowners, "I have bad news," and then after a commercial break is seen ripping off a piece of the wall to uncover the dreaded knob-and-tube system. This archaic method for transmitting electricity inside a house is a potential fire hazard, such that when upgrades are made to a house, the knob and tube has to be replaced if it is deemed unsafe (or possibly if the plot of the show is deemed to be lacking in dramatic tension).

Finding electric problems is a bit like debugging, and clearly, since knob and tube has been obsolete for seventy-five years, it falls in the category of a bug in somebody else's work. The difference is that knob and tube, despite being hidden behind a wall, is easy to locate: start with the plug on the wall and track it back from there. When I was called in to debug this mysterious problem in our software, I had no idea what I was looking for. Was it knobs? Was it tubes? Which metaphoric wall was I supposed to look behind? And if I found the right place, would the problem occur while I was watching?

I'll give away the ending: after a couple days of excavation, I found the bug and was rewarded with a bottle of champagne as well as the respect of my peers and a blissful return to more mundane tasks—until the next time I had to track down a flaky bug in somebody else's program. But let's take a detour to consider exactly how programmers approach writing and debugging software.

We will need something to debug, so below are two lines from a program in C#. The purpose of this code is to show the user an error message stored in a variable named ErrorMessage, of type string. This is a snippet of a program, so we assign a specific value to ErrorMessage (the text string "This is my error message"), as opposed to having it determined by an actual error:

```
string ErrorMessage = "This is my error message";
MessageBox.Show(ErrorMessage);
```

Look past the slightly backward syntax and arbitrary-looking punctuation; you may correctly infer that this code will cause the computer to show a message box—one of those pop-up windows that hovers over the screen—containing the text of the error message, as contained in the variable ErrorMessage. On my Windows 10 system, the message box—not the prettiest of message boxes, but you can see the connection between the code and result—looks like this:[1]

MessageBox.Show is an API, similar to what we saw in Fortran and BASIC, although in the argot of C# it is also known as a *method*. You might say, "My code calls the MessageBox.Show method." Or since programmers tend to talk of their code as an extension of themselves, you could assert, "I call the MessageBox.Show method." Or most commonly, since code that doesn't work yields the richest bounty of conversational fodder, you would be complaining, "I call the `MessageBox.Show` method and I can't figure out why it isn't working properly."

Given that software is built up in layers, this code is at a layer above `MessageBox.Show`, calling down into it. A similar concept exists in the physical world—the roof of a house may be built on prefabricated girders, which are themselves made of wood, steel, and nails. The electric appliances in your house are layered above the electric system in your walls, relying on it to provide power when needed. But these layers don't go that deep; the knobs and tubes are only one layer removed from what all visible to the homeowner. Some quick work with a claw hammer and all is laid bare for the camera. The `MessageBox.Show` method contains its own code, which in turn calls other code in a stack that is dozens of levels deep. Which means the knobs and tubes may be buried so deeply that you will never discover them until your software metaphorically catches fire.

In the code snippet above, `ErrorMessage` is passed as a parameter to `MessageBox.Show`. In C#, as in many modern programming languages, parameters are specified in a comma-separated list enclosed in parentheses, so we'll follow the convention that C# method names are followed with empty opening and closing parentheses in order to distinguish them from other programming constructs. `MessageBox.Show` will henceforth be styled as `MessageBox.Show()`.

The code in a program generally involves figuring out the desired parameters to a method, calling that method, and using the information returned from that method to decide what to do next. As code gets more complicated, the number of layers increases, and method calls are the glue that holds those layers together. A lot of code examples show only one layer, but that is unusual in real programs; code rarely proceeds for more than five lines without calling a method.

Let's change our code to display the error message in uppercase so as to be extra memorable. Uppercasing is easy using a method named `ToUpper()`; in C#, you call a method *on* a variable—in this case, `ErrorMessage`—by using *dot notation*, as shown below. We'll also add another string variable named `EM_Upper` to hold the uppercased string:

```
string ErrorMessage = "This is my error message";
string EM_Upper = ErrorMessage.ToUpper();
MessageBox.Show(EM_Upper);
```

The second line sets `EM_Upper` to hold the result of calling the `ErrorMessage.ToUpper()` method, and we then pass that as a parameter to `MessageBox.Show()`, instead of the original `ErrorMessage`. The error message is now displayed in uppercase, like this:

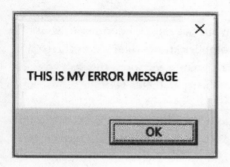

Now we add one more twist to this code: in addition to having the error message displayed in the message box, we will show a title at the top. We'll use the title "ERROR!," which we pass as a second parameter to `Message-Box.Show()`:

```
string ErrorMessage = "This is my error message";
string EM_Upper = ErrorMessage.ToUpper();
MessageBox.Show(EM_Upper, "ERROR!");
```

The difference is in the third line of code; `MessageBox.Show()` now has two parameters, separated by a comma. The string "ERROR!" is displayed as the title of the message box, where previously there had been no title:

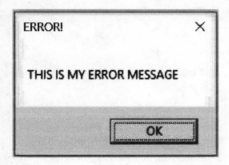

Both those parameters to `MessageBox.Show()` are of type `string` (the second parameter is not the name of a variable but instead contains the actual text of the string, surrounded by double quotes). There are multiple ways you can call `MessageBox.Show()` with different sets of parameters, because the people who wrote `MessageBox.Show()` (in this case at Microsoft, which invented C# and wrote the collection of methods that allow C# programs to do things such as show message boxes) decided it would be useful to offer the choice. The C# compiler knows the type of the parameters and their order, which form a signature of sorts; this call to `MessageBox.Show()` has the signature "first parameter is a `string`, second parameter is a `string`."

What if you accidentally got those backward, and specified the first parameter as the title of the message box, and the second as the text? Here is an example:

```
string ErrorMessage = "This is my error message";
string EM_Upper = ErrorMessage.ToUpper();
MessageBox.Show("ERROR!", EM_Upper);
```

It seems obvious to us what the intent is since we have been thinking about the code, and know that "ERROR!" is the title and the message is in `EM_Upper`. But the compiler doesn't know that, since the call still matches the method signature; one string is treated like any other string, and it lets the incorrect code through unchallenged:

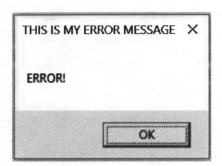

Fred Brooks, who managed both hardware and software teams at IBM, and later founded the Department of Computer Science at the University of North Carolina, once compared the creativity in programming to that in poetry, but noted, "The program construct, unlike the poet's words, is real in the sense that it moves and works, producing visible outputs separate from the construct itself. ... The magic of myth and legend has come true in our time. One types the correct incantation on a keyboard, and a display screen comes to life, showing things that never were nor could be." But he cautioned, "One must perform it perfectly. The computer resembles the magic of legend in that respect, too. If one character, one pause, of the incantation is not strictly in proper form, the magic doesn't work. Human beings are not accustomed to being perfect, and few areas of human activity require it."[2]

The things that make us swear at computers often involve them perfectly executing a set of instructions that don't do what we want them to do. In his book *I Sing the Body Electronic*, which chronicles a year embedded with a Microsoft development team in the 1990s, longtime Seattle observer Fred Moody recounts a discussion with a programmer:

Developers like to highlight the difference between the world of computing and the world outside of computing by citing the common directions on a bottle of shampoo—"Lather. Rinse. Repeat." ... In the everyday world, common sense tells you not to keep lathering and rinsing forever. In the world of computing, where there is no common sense and where everything must be rigorously defined, such an instruction is careless and dangerous.[3]

The joke is that the instructions don't tell you when to *stop* repeating; a computer would keep doing this until it ran out of shampoo (or possibly longer, if it neglected to stop when the shampoo bottle was empty).

Reversing two parameters to a method is the sort of mistake that can be hard for a programmer to catch, because our human commonsense filter kicks in when reading the code—the same common sense that makes us interpret the shampoo directions the way they were intended. The correct code looks maddeningly similar to the incorrect version; you see the title and message being passed to `MessageBox.Show()`, so what could be wrong?

Luckily there is documentation available for `MessageBox.Show()` explaining which parameter goes where, and even if you don't read the documentation and get the parameters backward, the error will be apparent if you run the program. In this case, you can quickly solve the problem: you know where the call to `MessageBox.Show()` is in your code, it's pretty obvious that you switched the two parameters, and presto, it is fixed.

Now imagine that the method you were calling was written by another programmer at your company and did not have clear documentation—a much more typical situation when debugging. Or imagine that the mistake did not result in a visually obvious, and otherwise harmless, transposition of two strings.

Although you can fix this problem by changing *your* code, you can't pin the blame squarely on your code or on the code in `MessageBox.Show()`; you just had a different understanding of how the parameters worked. The actual problem involves an interaction between your code and the details of how `MessageBox.Show()` orders its parameters; the bug is somewhere in the gap between them.

I want to emphasize this point: method calls hold together all the layers of software, and miscommunication across these layers is a major cause of unexpected problems. A collection of methods at a layer boundary is an API; I'll use that term to refer in an abstract sense to both a single method and collection of methods.

So to restate: APIs hold together all the layers of software, and miscommunication across these layers is a major cause of unexpected problems. This miscommunication can take a variety of forms, from not realizing the proper value that the API expects in a parameter, to not realizing how the API will interpret that parameter in guiding its internal logic, to misunderstanding when and in what form the API returns values. Many debugging sessions end with a programmer, after having examined their own code with a fine-tooth comb and found nothing amiss, cracking open the documentation, slapping their forehead, and exclaiming, "Oh, I didn't realize that the API that I was calling worked that way."

The decisions here are in the hands of the programmer writing the code inside the API; the caller of the API is stuck with whatever that other person decided. Often you cannot see the code in the API you are calling; it is provided to you only as compiled code, with no source code available. Unfortunately, people who write code that supplies APIs to other people generally don't spend a lot of time worrying about the external clarity of their API but instead get bogged down in the internal implementation of the API. This is because for any given code you want to write, there are likely multiple ways to write it, and not a lot of wisdom about which one to choose.

The book *The Paradox of Choice*, by the psychologist Barry Schwartz, argues (in a section of the book titled "Why We Suffer") that having more choices does not make people happier:

Freedom and autonomy are critical to our well-being, and choice is critical to free-dom and autonomy. Nonetheless, though modern Americans have more choice than any group of people ever has had before, and thus, presumably, more freedom and autonomy, we don't seem to be benefiting from it psychologically.[4]

He then goes on to explain that too much choice can sometimes be a burden on us. Programmers are frequently victims of this; there are so many ways to write even trivial code like the stuff above, it is hard to know what the "right" way is.

Getting back to our code from before, here is the correct version (with-out the swapped parameters to `MessageBox.Show()`) as we last saw it:

```
string ErrorMessage = "This is my error message";
string EM_Upper = ErrorMessage.ToUpper();
MessageBox.Show(EM_Upper, "ERROR!");
```

Observe that in the second line, we declare a new variable, `EM_Upper`, to hold the uppercase version. This seems reasonable; although we don't need the original mixed-case error message, we may want it at some point in the future, so we allocate a second variable to hold the uppercase version and keep the original in `ErrorMessage`.

Hang on. Do we need to retain the original value of `ErrorMessage`? Sure, we might need it in a future version of this code, but right now we don't. And if we don't need it, we can avoid `EM_Upper` entirely and instead replace `ErrorMessage` with its uppercase version and pass *that* to `Mes-sageBox.Show()`:

```
string ErrorMessage = "This is my error message";
ErrorMessage = ErrorMessage.ToUpper();
MessageBox.Show(ErrorMessage, "ERROR!");
```

Come to think of it, since all we do with the uppercase error message is pass it to `MessageBox.Show()`—we don't use the original *or* the uppercase version in any later code—we can combine the last two lines of code into one:

```
string ErrorMessage = "This is my error message";
MessageBox.Show(ErrorMessage.ToUpper(), "ERROR!");
```

Except—and I'm just pointing this out to be helpful, you understand—now we haven't saved the uppercase version anywhere. We passed it to `MessageBox.Show()` and then it disappears into the ether. Does this matter? If the code later needs the uppercase value a second time, we would need to call `ToUpper()` again, which seems wasteful, so shouldn't we save the output of `ToUpper()` somewhere just in case? And if we do decide to save it, should we stash it back in `ErrorMessage` or create `EM_Upper` as a separate variable, thus retaining the original version as well but using a little more of the computer's memory?

These last few paragraphs of discussion are all theoretical. The code works right now, so why complicate it? Why are you worrying about saving both versions *or* calling `ToUpper()` twice when the code needs to do neither of those things?

Sure, replies the devil on the other shoulder, but maybe if you set yourself up for the future now, when you are familiar with the code, you will have less chance of making a mistake later and save time overall. A well-prepared devil might point out that one study of software mainte-nance noted, "There is a unique maintenance aspect called 'knowledge recovery' or 'program understanding.' It becomes a major cost compo-nent as software ages (assume 50% of both enhancements and defect fix-ing)."[5] Half your future maintenance costs will be spent relearning the details of your program that you will have forgotten in the meantime! Surely it is better to make those changes now, when the code is fresh in your mind.

Keep in mind we're talking about three lines of code here.

This problem is not unique to software; many tasks can be accomplished in multiple ways. What's different is the ease with which you can change your mind and update your code, and the lack of criteria for determining which approach will be most useful in the long term. If you are building a bridge, you will know the distance it is supposed to span and weight it is supposed to hold, and it is understood that the current design is based on those factors. Nobody is going to assume that the same bridge design, with just a few modifications, will handle twice as much distance or weight, or that it will be simple to accommodate such a change when the bridge is halfway built. Changing software is so easy—a few keystrokes, a wave of the compiler, and you are done—that the temptation is always there, and the incentive to figure everything out ahead of time is less, for the same reason. The end result is that almost every piece of software written eventu-ally winds up being modified to solve a different problem than what it was designed for.

"The computer's flexibility is unique," points out the researcher John Shore in his essay "Myths of Correctness":

No other kind of machine can be changed so much without physical modifications. Moreover, drastic modifications are as easy to make as minor ones, which is unfortunate, since drastic modifications are more likely to cause problems. With other kinds of machines, drastic modifications are correspondingly harder to make then minor ones. This fact provides natural constraints to modification that are absent in the case of computer software.[6]

André van der Hoek and Marian Petre, editors of the book *Software Designers in Action*, observe, "Almost any product can be changed, in some way, after it is delivered. What makes software unusual is the expectation of the customer, the user, and other stakeholders, that it *will* change."[7] Since software almost always has a potential alternate future ahead of it, there is no statute of limitations on suggested improvements. And you never know, unless you can see the future, which ones will result in real savings and which ones will be needless complications.

While the choices we're discussing here—what API to call, or whether to use a variable or not—seem innocuous in this situation, these are precisely the sort of choices that can, if made incorrectly, lead to software that has real problems, that crashes, hangs, or allows other users to steal your files. The electricians who installed knob-and-tube wiring back in the 1930s were following correct, state-of-the-art practices for the time; it was only determined in retrospect that they were actually creating a large, expensive, and potentially life-threatening problem for future generations to deal with. Is that API choice you are making a clever decision, or will it be determined by a future programmer, as they slander your name, to be horribly misguided?

Still, you can't dither forever, so eventually you choose a spot on the continuum of present versus future gratification, and write your code to match. Are you done now?

No, unless you are programming all by yourself, which usually only happens during the sunny idyll known as college (or high school). You are now a professional programmer, working with other programmers, so what comes next is an opportunity for all your coworkers to give their opinions through an activity with the seemingly auspicious name of *code review*.

A code review is where other people offer constructive criticism of your code. This sounds like a good idea, like grabbing a second electrician to give your work the once-over. That's what you would expect people to do as they graduate from do-it-yourself wiring projects in their own home to becoming professional electricians: find somebody else to point out your knobs and tubes before they can cause any fires! The language used gives this analogy a helpful shove in the wrong direction, because electricians are governed by an electric code (that word again). The electric code is where it is written down, "Thou shalt not use knob-and-tube wiring for new homes," and more important, "When evaluating existing knob-and-tube wiring, these are the things you look for to determine if it is safe." The phrase "code review" implies that your fellow programmers are noting deviations from accepted practices and standards, comparing your code to a "professional programmer's [the other kind of] code."

Unfortunately there is no equivalent of an electric code for software, and the books available to programmers, although they offer advice in some of these areas, are not backed up by any empirical studies; they tend to add fuel to both sides of a debate. A code review is really about other programmers giving their personal opinions on how they would have written the code, backed up by nothing more than their own experiences. And since your peers know all about how flexible code is, they are likely to feel that their suggestions should be adopted, no matter how late in the game it is, because the game never ends. This makes them more likely to suggest changes and more likely to pooh-pooh your code if you don't agree with them.

The most likely feedback from a code review is other programmers' opinions on the same questions that you noodled over before you even sent the code out for review: Should you modify your code now in anticipation of future needs, of which you are currently unaware?

Consider variable names—a favorite topic. Our code above has two variables, ErrorMessage and EM_Upper. Those names are inconsistent; the second one has an explanation of the meaning of the variable (it's in uppercase) that is absent from the first one, while the first one spells out the purpose of the variable (it's an error message), but the second one uses initials. We, the author of the code, know how it evolved to this point: we started with only one variable, ErrorMessage, and added the second one later. Until we added the second one, we didn't know what was going to distinguish it from the first, so plain-old ErrorMessage seemed reasonable. Meanwhile, when inventing EM_Upper, we decided to save a bit of typing and shorten the first part.

Now once we did add the second variable and the case—mixed versus upper—became the distinguishing factor, we could have renamed all uses of the original `ErrorMessage` to be `ErrorMessageMixed`, but really we should then go change `EM_Upper` to `ErrorMessageUpper`, or alternately change `ErrorMessageMixed` to `EM_Mixed`. And come to think of it, the initial error message might already be in uppercase; we can't assume it is mixed case, so perhaps `ErrorMessageOriginalCase` would be a better name. Which would mean `ErrorMessageUpper` should become `ErrorMessageUpperCase`. Meanwhile we're still not 100 percent sure we need two variables, so are we willing to commit to all that typing? Looming over this is the nonzero chance of making a mistake (in particular, accidentally replacing one of the uses of `ErrorMessage` with `ErrorMessageUpper` instead of `ErrorMessageMixed`, which would compile fine, but botch everything when you ran it).

A second programmer arriving on the scene for a code review knows none of this history, nor do they appreciate your inner struggle. What they see is the inconsistency between the names `ErrorMessage` and `EM_Upper`, which they will likely point out. By the way, variable names go away when the program is compiled, so changing the variable name has no effect on how the program runs or the user's experience running it. This is just programmers arguing about readability for the sake of future programmers who encounter the code.

Believe it or not, one of the most contentious questions, which you can't see at all when reading code in a book, is this: When indenting code, which happens a fair bit, do you type a series of spaces or a single tab character? Some people like to see code indented four spaces each time, and some like to see it indented eight spaces; using tab characters means each person can see the indent level they like by adjusting the tab settings on their own computer, but some people consider that heresy and think the original programmer should be able to control exactly how the indenting is seen. Worse, a file with a *mix* of tabs and spaces can devolve into a visual torment. When I worked on the first versions of Windows NT (the precursor to today's Windows) back in the early 1990s, there was a strict rule that there would be no tab characters in the source code, on pain of baleful stares from your coworkers (once they got finished removing the tabs from your code). To this day, if you want to see an eye roll from a programmer, utter the magic phrase "tabs versus spaces."[8]

So you've got the variable names and indenting, and further discussion topics such as whether you should put a space before the equal sign when assigning to a variable, before the opening parenthesis of a method call, after the comma that separates method parameters, before the semicolon at the end of a line, or really anywhere there is or isn't a space (which is almost anywhere, since most programming languages ignore extra spaces in the code), or whether blank lines in your code are a sinful waste or glorious luxury.

Harlan Mills once wrote about a similar situation, maintaining, "Since there was no mathematical rigor to inhibit these discussions, some became quite vehement."[9] Vehement is an understatement; these arguments are often called "religious" because they rely entirely on faith, not demonstrated evidence.

Despite the energy expended, code reviews rarely turn up real user-visible bugs. They are more about enforcing local norms such as "this is how we name our variables." Really bad bugs, security issues, or potential crashes usually involve a series of mistakes, each small enough to go unremarked in isolation, acting in unfortunate concert with just the wrong set of data. Oftentimes they are a misunderstanding between the programmers responsible for two adjacent layers of code at the API boundary. Brooks recognized this problem forty years ago: "The most pernicious and subtle bugs are system bugs arising from mismatched assumptions made by the authors of various components" (the loosely defined terms *component* and *module* are often used to denote "a bundle of code that provides a set of related APIs").[10] Code reviewers do try to anticipate future modifications to the code, but when reviewing code that provides an API, they rarely try to predict future misunderstandings by callers of the API, especially since that code hasn't been written yet. When your perspective is from the inside of an API upward, it all seems perfectly logical.

As a result, code reviews almost never get into the issue of the usability and clarity of the APIs being exported to a layer above. The API name and parameter list is the box that holds the code being reviewed, normally accepted as fact while your eye slides over it to get to the meaty algorithmic parts inside. In a way that makes sense, since the internals are the "hidden" part that may never get looked at again, while the API external surface will be seen by any other programmers who call it, but the latter will have a disproportionate effect on whether code from two different programmers will work together as planned. And once the code providing an API is judged complete, programmers are even less likely to change the API name and parameters than they are to rename an unclear variable; for one thing, they would also need to change any code that is now calling that API.

Having said that, I do applaud code reviewers for worrying about the readability and maintainability of the code, because while it may not have much effect on clarity across API calls, it relates directly to another important source of bugs, which is code handoff across time: the situation where another programmer needs to modify the code for future use (or where you, as mentioned earlier, come back to the code some time later; the amount of time it takes your own code to become foreign to you is depressingly short). The code reviewer is a good simulation of that future person since they themselves do not know the code either. The problem is that while the code reviews are well intentioned, it's not at all clear that the problems they point out will have an effect on future maintainability; it's frequently a "he said, she said" sort of argument (or unfortunately, often a "he said, he said" one).

At one point in Microsoft's history, there was a push to write code using *Hungarian notation*, where variable names were prepended with little duodenums that described their type—such as a number, string, and so on (the "and so on" exists because programmers can create their own types, built up from collections of strings and numbers). For example, all variables that were strings would start with sz, so Hungarian-styled code was peppered with variable names like szUsername and szAddress. The theory was that it was useful to know at a glance, without knowing anything else about how the code worked, if a particular variable was a string or number, to prevent you from accidentally using one where the other was called for. This led to extreme all-sizzle-no-steak examples like szA, where sz informed the masses that this was a string, but the A part was a flashback to the days of BASIC, which did nothing to tell you what the string was used for.

Hungarian was a source of contention in the halls of Microsoft; the Office team adopted it, but the Windows NT team thought it was silly, so I thought it was silly by osmosis (would the discussion above, about EM_Upper versus ErrorMessageUpperCase, have been meaningfully impacted if it instead were about szEM_Upper versus szErrorMessageUpperCase?). The Windows NT naming style tended toward long intercapped names, such as MaximumBufferLength, a style known as "camel casing" because the capital letters look like a multihumped Dr. Seuss camel; the name told you a lot about the purpose of the variable but kept mum on its type.

(It will be important to certain readers for me to clarify that camel case actually has the initial letter in lowercase, as in `exampleVariableName`, and the ones with the first letter capitalized that we used in Windows NT were called "Pascal case," but camel case is a much more evocative phrase.[11] Anyway, back to our story.)

Proponents of Hungarian in turn derided these long, not-quite-camel-case names as being error prone as well as wasteful of keystrokes and disk storage (two things that had historically been in short supply for programmers, although truthfully no longer were at that moment in history). Which led to the counterargument that it is clear enough that the variable `CurrentlyLoggedInUserName` is a string and no extra characters at the beginning are needed to indicate that. In addition, by the early 1990s compilers were better at recognizing when code was passing the wrong type of variable around (enforcing that your method call matched the method signature, such as not using a string when a number was called for, was an innovation in compiler technology whose absence had in the past caused all sorts of entertaining bugs). This made Hungarian less necessary than it was in, say, 1986: if the compiler is going to catch a type mismatch, then you don't need Hungarian; and the more interesting mistakes do not involve bollixing up strings and number but are instead about using one more complicated type where a different more complicated type is needed, and those complicated types won't have easily recognized Hungarian prefixes to guide you.

Since the Office and Windows NT teams had their own separate piles of source code, pro- and anti-Hungarian arguments could be lobbed back and forth with no ground given; each side had the other's worst-case offenders to parade around the public square, with camel casers chuckling at `pwszA` and Hungarian advocates snorting at `SomeReallyLongVariableName`. If there was ever a thought of compromise (how about long camel-cased names with Hungarian prefixes also?), I never heard about it. This was serious business, with no time for such foolish ideas! Besides, anything other than complete capitulation by the other side would have meant fixing up a lot of variable names in your own code—a daunting prospect that nobody wanted to tackle. The few programmers who were brave enough to switch teams were quickly assimilated into their new culture, and the twain never met.

To add fuel to this bonfire of whataboutism, the two sides weren't arguing about the same thing. The original Hungarian system, which became known as Apps Hungarian because it was used in the division that wrote applications such as Office, prescribed prefixes that were more informative than just the type of a variable; you might distinguish a variable that held a row number from one that held a column number by using the prefix `row` or `col`.[12] Somehow (the blame is generally placed on the team that wrote the documentation for the Windows API, apparently following a misguided impulse to simplify the notation) it made its public debut in a form known as Systems Hungarian, in which the variable name prefixes only identified the type—as in number versus string—which is much less useful (although the more your Hungarian prefixes resemble real words, the more the difference between Hungarian and Windows-NT-style boils down to the capitalization of the first letter—still fertile ground for religious argument, of course).[13] Thus the Apps Hungarian that was venerated by the Office team was different from the Systems Hungarian that was used as a punching bag by the Windows team.

By good fortune, a writer named G. Pascal Zachary wrote a book about that Windows NT project and recorded his impressions of the Hungarian battle raging in Redmond, Washington, at the time:

Some disputes, however, involved what programmers call "religious differences." The points at stake seem important only to zealots; a neutral party might say that both sides are right. But zealots—unable to silence their opponents with logical arguments—hurl insults.

One of the oddest disputes, which brought out the worst in zealots, involved the notational system used to write instructions in C, one of the most popular computer languages. Over the years Microsoft had adopted its own conventions, called Hungarian, after its creator, Budapest-born Charles Simonyi. ... [I]t lacked the ready familiarity of conventional notation, which relied largely on English words rather than opaque abbreviations. The differences between the two styles spawned many arguments, whose merits were lost on outsiders.[14]

One of the programmers on the team (not me) is quoted describing Hungarian as "the stupidest thing I'd ever seen," although it's unclear if he is referring to Apps Hungarian or Systems Hungarian. He adds the quasiwise summary, "Coding style wars are a waste of valuable resources, although the confusion caused by Hungarian probably wastes more time."[15] And if his arguments sound reasonable, remember he was part of the crew that obsessed over tabs versus spaces for indenting source code. The same programmer can have a perfectly rational, "live and let live" attitude about, say, spaces between method parameters, but go into conniptions at the sight of an unneeded blank line in the source code. For that matter, I wouldn't have described the variable names used in Windows NT as "conventional notation"; they seemed oddly long to me when I first joined the team, being used to "opaque abbreviations" sans Hungarian prefixes.

Luckily the code reviewers will eventually stop commenting, or you will get tired of listening, and you can update your code to reflect the feedback you choose to heed. Of course, any change to your code is an opportunity to make new mistakes that will in turn require their own debugging; it is a particularly numbing experience to decide that today is the day you are going to do your civic duty and rename that obscure variable, only to discover that you have accidentally broken something while making the change, and the compiler is now complaining that "an expression tree lambda may not contain a coalescing operator with a null literal left-hand side"—an actual C# compiler error, albeit one that is unlikely to be caused by a typo in a variable name.[16]

Let's make one more change to our code: have it only display the error message if that message contains the word "JavaScript" in it (JavaScript is another programming language). Since we have uppercased the message, we can check if it contains the capital word "JAVASCRIPT" using the Contains() method (I've removed the first line, where it explicitly sets ErrorMessage to "This is my error message," because it would make the code look slightly ridiculous; clearly that string does not contain "JAVASCRIPT," so there's no reason to check):

```
string EM_Upper = ErrorMessage.ToUpper();
if (EM_Upper.Contains("JAVASCRIPT")) {
    MessageBox.Show(EM_Upper, "ERROR!");
}
```

The line that reads

```
if (EM_Upper.Contains("JAVASCRIPT")) {
```

performs a test; if EM_Upper contains the string "JAVASCRIPT" anywhere within itself, then the code between the { } runs, and otherwise it doesn't. The word if is a recognized keyword in the C# language; for notational convenience, I am going to write such keywords in capital letters, even in languages that traditionally are written in lowercase, so it will be referred to as an IF statement.

As conscientious programmers, we are aware that our code can run in multiple countries, where the error messages might be translated into a different language, but we have been assured that the term JavaScript, being the name of a programming language, won't be translated.

Is this correct? Well, the basic idea is correct, but it does have a bug, and you might not realize that for a while, because it depends on a detail of the implementation of a method that you—and many experienced programmers—are likely completely unaware of.

As an example of nonobvious method implementation details, consider the internals of MessageBox.Show(), the actual code that shows a message box. An important aspect of writing the code was deciding what behavior made sense to callers of their method and how that behavior should be accessible via the parameters.

You may have noticed, from the screenshots earlier in this chapter, that in addition to showing the message and title, the message box will display a button labeled "OK" that the user can click:

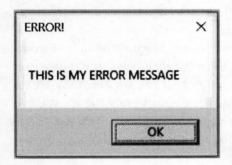

This is perfectly reasonable, noted in the documentation, and apparent when you run the program, but not clear from knowing the method name and parameter list.

The situation appears benign: MessageBox.Show() will show a button that the caller didn't explicitly request, but is that harmful? The answer is "no" in this case. In fact, by passing extra parameters, you can have some control over what buttons are shown—assuming you know that the method supports this. Yet methods having unknown side effects is the cause of many bugs.

That was exactly the situation with the bug at my first job out of college, where the street address of certain doctors was being replaced. When I was enlisted to help, it was quickly apparent that the problem was in the API that retrieved the doctor's information from the database on the computer (as opposed to, say, retrieving the right address but messing up the code to display it). The exact name of that API escapes my memory, but it doesn't matter; we'll call it `GetDatabaseRecord()`. This API presumably took parameters specifying which doctor to retrieve, although those don't matter either. Of course `GetDatabaseRecord()` was itself built on other API calls, which were built on other API calls, and so on. My task was to paw through these underlying layers of code and find out why they were occasionally misbehaving.

After some investigation, I discovered that another programmer had modified a section of the program to calculate and display extra data about the doctors in the database. In certain cases, this required them to load other doctor records out of the database (I don't recall the details, but let's say that in the case where two doctors had attended medical school together, the database had this noted somewhere—something that would only be true in certain instances, and didn't jump out as an obvious difference between the doctors showing a corrupted address and the others, so it was not noted in the repro steps). Because the street address of a doctor, somewhat uniquely among all the other data fields in the database, was a string that could be of wildly varying length, we stored the "street address of the last doctor loaded from the database" in a specific variable in memory that had enough room for any reasonably sized address. This is the sort of optimization you made to save memory on the underpowered computers of the day.

In the feature that the other programmer was adding, the street address wasn't needed, so this change didn't matter, but it meant that sometimes the address in that special "street address of the last doctor we loaded from the database" variable wasn't what we thought it was, because while loading information for one doctor, we proceeded to load information for his medical school buddy, and this updated the "street address of the last doctor we loaded from the database" variable. This was all happening several layers below the code that called `GetDatabaseRecord()`; that code hadn't changed, but the internal behavior of `GetDatabaseRecord()`—or more precisely and vexingly, the internal behavior of an API that was itself called several layers below `GetDatabaseRecord()`—had changed. This wasn't maliciousness on the part of the other programmer, yet it was subtle enough that even she herself, when the "wrong address" problem was first being looked into, didn't realize that her earlier change was causing the problem.

Once again you are at the mercy of whoever wrote the API you are calling—not only to define the parameters in a logical order, but also to document all the intended side effects, even if it isn't obvious why they would matter. When calling an API, you have precious little information about the details of its implementation and how reliable it is—just a name and a parameter list, as a thin line of glue holding together your software.

I don't remember exactly how I *fixed* the problem, but it was easy once I had found it; the solution can be left as an exercise for the reader. I might have added an extra parameter to `GetDatabaseRecord()` to tell it "don't load extra doctor information" and made that parameter "true" in this specific case. If I was feeling motivated, I could have rewritten the code that used the single "street address of the last doctor we loaded from the database" variable so that it instead used multiple variables as needed. This second way would have been more "correct," but it also would have been a larger change, delaying my champagne reward, with more risk of breaking something else while making the fix. On the plus side, it would have meant that a future caller of the `GetDatabaseRecord()` API had less to understand about the implementation, which would make things less error prone. Just as people argue, during code reviews, about the correct way to write a piece of code, they also argue about the correct way to fix a bug once the cause is found, typically involving this sort of trade-off between "less immediately risky but somewhat ugly" and "more complicated but more elegant for the long term."

Let's return, finally, to our code that checks if the error message contains "JAVASCRIPT," which I claimed a few pages ago had a real bug in it: a monolingual English speaker may be unaware that the concept of uppercasing is subject to regional interpretation. English has a lowercase i with a dot on top and an uppercase I with no dot on it; for your reference, I have included several examples of both in this sentence. Most languages written in the same alphabet have the same lowercase i and uppercase I. In Turkish, however, there is a lowercase dotted i and uppercase dotted İ, and a lowercase undotted *ı* and uppercase undotted *I*. And the dotted-or-not aspect doesn't change when you capitalize, so the capital of *i* is *İ*, not *I* as it is in English. When you uppercase a word with an *i* in it, such as the string "JavaScript," the capital in English is "JAVASCRIPT" and the capital in Turkish is "JAVASCRİPT" (notice there is a dot above the uppercase *I*). And despite what a common sense–laden human might think, to a computer those are most definitely not the same thing.

If you call `ToUpper()` with no parameters, as we did, the implementation uppercases based on the language setting configured on the computer, which the user can choose. If your user is on a computer configured for Turkish, the uppercasing of "JavaScript" will be different than if the machine is configured for English, and the `Contains()` method won't match it as expected. You might have thought it was clever to have your code make the comparison against the uppercased version of the error message, but this could cause hard-to-diagnose problems, especially if you try to reproduce the bug on a machine configured for English.

The fix to our Turkish uppercasing problem is simple once you know about it; change your `ToUpper()` call from[17]

```
EM_Upper = ErrorMessage.ToUpper();
```

to

```
EM_Upper = ErrorMessage.ToUpper(InvariantCulture);
```

As with the `MessageBox.Show()` method, `ToUpper()` has multiple versions that take different parameters. The simplest version assumes that it should use the *culture* (the preferred term since it encompasses more than just language, extending to areas such as currency symbols) that the computer is configured for. This is normally right, but not in the case of comparing uppercased strings; passing `InvariantCulture` as a parameter to `ToUpper()` tells it to uppercase in a way that is guaranteed to be the same on all computers (here's an insider tip: the secret is "always do it like they do in English"; it doesn't have to be politically correct, just consistent).

This works fine, but just like wanting `MessageBox.Show()` to display something other than an "OK" button, you have to know to do it. The parameterless "use current culture" version of `ToUpper()` is more convenient to call (with convenience defined as less typing by the programmer) and easier to discover, but its existence allows the calling code to be unaware of the notion of cultural differences in uppercasing, which is bad. The problem is not just that the default local versus invariant culture choice made by ToUpper() is a hidden choice made by somebody else; it's that the *existence* of such a choice might be unknown to the caller, either because a programmer doesn't know about cultures at all or they don't realize that the "tell me which culture to use" version exists.

One of the tricky aspects of designing a method is deciding what to make a required parameter (which always has to be passed in, such as the title of the message box or string you want to uppercase) versus an optional parameter (the buttons to display in a message box or culture to use when uppercasing), the related question of what the default behavior should be if the optional parameters are not specified, and lastly, what is not even specified via a parameter (such as the font to use in a message box) and therefore gives the caller no choice at all.

As usual there is no one right answer, despite many brain waves having been expended on code reviews of these questions. Whatever decisions are made, ignorance of the default behavior of a method is a common problem. Viewed through that lens, the existence of the default-culture version of ToUpper() is not a convenience but rather a tragic mistake, source of needless bugs, and missed opportunity to educate programmers about regional differences—all for the sake of saving a little bit of typing! Default parameters, while generally considered a useful convenience, likely do more harm than good. They are essentially like allowing a bridge builder to say, "Give me some steel to build my bridge," rather than requiring them to always specify the exact properties of the steel that they need. Programmers unknowingly think the simplest call is the right one—until they get an irate call from Ankara.

Brooks explained the difference between a program—"complete in itself, ready to be run by the author on the system on which it was developed"— and a programming systems product—"the intended product of most systems programming efforts."[18] To get from the former to the latter, you introduce complications in two dimensions. The first complication is going from a single author to a program that can be "run, tested, repaired, and extended by anybody." The second complication is going from a single program to "a collection of interacting programs, coordinated in function and disciplined in format, so that the assemblage constitutes an entire facility for large tasks."[19]

Back in high school and college, I was working on plain-old programs, but in industry I was working on programming systems products. As Brooks noted, the two big new complications in making this transition are communication between programmers across time and communication between components across API boundaries.[20] These are two areas where my self-taught education had left me severely lacking.

Brooks added that "this then is programming, both a tar pit in which many efforts have foundered and a creative activity with joys and woes all its own."[21] Don't programmers learn how to deal with these problems correctly? They may eventually. But given their self-taught beginnings, they tend to be focused on another aspect of their software, which I'll discuss in the next chapter.

4 The Thief in the Night

What do programmers from my era worry the most about? How efficiently their programs run. For a glimpse, we turn to the 1986 book *Programming Pearls*, a collection of columns that Jon Bentley, a former Carnegie Mellon computer science professor who worked at Bell Labs, wrote for the journal *Communications of the ACM*.

Bentley worked on UNIX, which ran on minicomputers that weren't as cramped as PCs.[1] Nonetheless, looking over the articles he chose, he comments that performance—making sure that programs run as quickly as possible and using as little memory as possible—is a theme that runs through all of them.[2] In fact, he fears that readers might be ignoring the importance of performance. In a chapter titled "Squeezing Space," Bentley writes, "If you're like several people I know, your first thought on reading the title is 'How old-fashioned!' In the bad old days of computing, so the story goes, programmers were constrained by small machines, but those days are long gone. The new philosophy is 'a megabyte here, a megabyte there, pretty soon you're talking about real memory.'"[3]

Coming from a UNIX environment, Bentley was concerned that the increased computing power of minicomputers, coupled with the relative lack of memory constraints, was going to lead to a generation of programmers who were oblivious to performance. He needn't have worried. More and more programmers were doing their work on personal computers, and his message found a receptive audience among them. PCs had grown rapidly in storage capacity since our original IBM PC and its 64 kilobytes of memory in 1982, but even the more advanced IBM PC AT, released in fall 1984, was limited to 16 megabytes of total memory, so a megabyte here or there really did matter.

Bentley was a performance guru; his 1982 book *Writing Efficient Programs* focused exclusively on the topic. Yet he was not a mindless performance improver. He cautioned that performance improvement techniques should be applied with care, and only when needed: "The rules that we will study increase efficiency by making changes to a program that often decrease program clarity, modularity, and robustness. When this coding style is applied indiscriminately throughout a large system (as it often has been), it usually increases efficiency slightly but leads to late software that is full of bugs and impossible to maintain."[4]

That subtlety was frequently lost amid the PC-inspired zeitgeist of the time. And although *Writing Efficient Programs* had examples in Pascal, and *Programming Pearls* used a mix of languages, most languages, even those that allowed reasonably "structured" programs—Pascal, PL/I, ALGOL, and newer versions of Fortran—were in the process of being swept aside by the language that I was first exposed to in my sophomore year of college—a language that had a strong focus on performance: C.

The C language was invented in the early 1970s at Bell Labs, the same place where Bentley wound up working. Bell Labs was the research arm of the Bell Telephone Company, which at the time had a monopoly on long-distance calling; it needed complicated software to handle the routing of phone calls. To support this, it wrote the operating system UNIX, originally in assembly language, as almost every operating system was back then. Higher-level languages were viewed as producing code that was too slow and bulky to run in the guts of an operating system. Yet when it came time to port UNIX to run on a new computer, the decision was made to rewrite it in a high-level language. No suitable language existed, which led to the birth of C.[5]

It is hard to describe how "right" C felt to someone like me, who was used to fighting memory limits on an IBM PC. In Pascal, blocks of code are delineated with the keywords BEGIN and END; in C, they use { and }. If a programming language construct could be described as aerodynamic, this was it. It's a small thing that has no effect on the code that the compiler produces, but it made C feel sleek and modern, while Pascal retained a faint air of tweed jackets and elbow patches.

The designers of C took great pains to ensure that the language did not insert any roadblocks to performance. The result is a language that takes care of the grunt work of mapping variable names to memory locations, handling loops and IF statements without requiring explicit GOTOs, and passing parameters to functions, without getting in the way of anything else. C has been characterized as a thin wrapper around assembly language.[6] This could be meant as both a compliment and insult, but it perfectly fit the needs of the transition from software running on mainframes to software running on PCs, once the complexity of that software got much beyond DONKEY.BAS.

As Bentley said, concentrating on performance "indiscriminately" can cause problems, but the design of C forces you to do so, because that's how the language works. One particular performance-focused feature of C is the worst of the bunch, and it requires a little background to explain.

An important way in which C was close to the processor was how it handled numbers of different *bitness*. Down at the level of machine/assembly language, you can control how many bytes are used to store a number. The smallest size possible is 1 byte, which is 8 bits,[7] with 1 bit being a single binary digit that is either 0 or 1 (if you are old enough to remember learning about bases in new math class, "binary" and "base 2" are the same thing). The lowest 8-bit number, written in binary, is

```
00000000
```

which is 0, and the highest 8-bit number, written in binary, is

```
11111111
```

which is 255. So an 8-bit number can store values from 0 to 255; you can also tell the processor (via how you encode any single machine language instruction) that it should treat the highest bit as a *sign bit*, indicating positive or negative, in which case you can store values from –128 to 127 instead of 0 to 255.[8] When 8-bit numbers are interpreted as ranging from 0 to 255, they are referred to as *unsigned*, and when they range from –128 to 127, they are referred to as *signed*.

Moving up to 2 bytes of storage gets you 16-bit numbers, ranging from 0 to 65,535 (or –32,768 to 32,767 for signed numbers); 32 bits gets you 0 to 4,294,967,295 (or –2,147,483,648 to 2,147,483,647); and 64 bits gets you 0 to over 18 quintillion—quite a large number (or a signed range of –9 quintillion to 9 quintillion, which are also large numbers).[9] When somebody talks about a 32- or 64-bit processor, they are referring to the largest number that the processor can deal with in a single machine language operation, which corresponds to the size of each register. A 32-bit computer has 32-bit registers and can add, subtract, and so on two 32-bit numbers (I'm glossing over things a bit, but not in a way that matters here).[10]

The processor can also operate on fewer bits at a time, so you can, for example, tell a 32-bit processor (again via the details of how you encode a machine language instruction) to only add, subtract, or move to/from memory 8-bit numbers. If you knew that a certain number would never go above 255, then you could store it in 8 bits and operate on it in 8 bits, which would result in slight savings in both memory and speed.

High-level languages before C had a single "integer" type that generally corresponded to the bitness of the processor, so the largest number a program could handle depended on the computer it was running on.[11] Furthermore, for simplicity they supported only signed numbers, since it was more likely, in the universe of programs being written at the time, that the user would want to store a small negative number than a large positive one. The original IBM PC was a 16-bit computer, so the INT (integer) type in BASIC was signed 16 bits and therefore supported values from –32,768 to 32,767. If you had two integer variables A and B that were both equal to 25,000, and you executed this BASIC command

```
10 D = A + B
```

then BASIC would realize that the value that D was supposed to have, 50,000, was too big to fit into a 16-bit signed integer and it would terminate with an "Overflow" error.[12] This is not ideal, but it's better than silently ignoring the overflow and proceeding as if nothing happened, with a bogus result in D (which for reasons that are complicated to explain in detail but conceptually involve the overflowed number "wrapping around," would have been the unexpected value of –15,536).

In C, by contrast, all this detail was exposed directly to the programmer; when you declared an integer variable, you indicated whether it was going to be an 8-, 16-, or 32-bit number (later extended to support 64-bit numbers).[13] Also, to allow you to fit in twice as many positive numbers if you knew a variable would never be negative, you could explicitly declare whether a variable was signed or unsigned, and the compiler would keep track and encode the appropriate machine language instructions.

Moreover, C did not check for overflow when performing mathematical operations on numbers. In order to check for overflow on

```
D = A + B
```

a compiler must generate code—that is, automatically include it in the machine language code that it creates to calculate that expression—that does this:

1. Calculate A + B
2. Check if that last operation overflowed (which is something the processor will have made a note of)
3. If it did overflow, print an error and terminate
4. If it did not overflow, store the result in D

Being performance focused, C skipped the second and third steps; it's one of those small things that add up if you do it every time you perform a mathematical calculation. C would ignore the potential overflow and store the result in D, no questions asked. In most cases this is fine; if a video game is storing the on-screen coordinates of an alien in a 16-bit integer and your screen is a thousand pixels wide, you don't need to worry about overflow as the alien moves across the screen.[14] On the other hand, numerical overflows can cause major bugs: one of the software problems with the Therac-25 radiation therapy machine, which killed three people in the 1980s due to overdoses of radiation, was a bug that hit when an 8-bit counter variable overflowed back around to 0 at just the wrong time, and an Ariane 5 rocket self-destructed in 1996 due to another overflow bug, trying to copy a 64-bit value into a 16-bit variable.[15]

The C language makes a similar optimization with the programming construct known as an array, which exists in almost all programming languages. An array lets you declare a single variable that can hold multiple values. For example, in Pascal you can declare a single integer this way:

```
var a: integer;
```

while you declare an array like this:

```
var a: array[0..4] of integer;
```

and you can then access the 5 elements of the array as a[0], a[1], a[2], a[3], and a[4], each of which can hold an integer value. Importantly, the *index* into the array, the part in the square brackets (also known as a *subscript*), can be a variable instead of a constant number, so you can write a loop like this:

```
for i:= 0 to 4 do
begin
```

```
    writeln(a[i]);
end
```

to print out the different elements of a (writeln() is the Pascal API to print out a value, and BEGIN and END, as noted before, demarcate blocks of code). If you try to access an element of the array a with an index of 5 or more, then the program will terminate with an error.

If the array index is in a variable, it means the compiler can't figure out, at the point when it is compiling the program, if an array access is going to be legal because it won't know what value the variable will have when that line of the program is executed. It has to add code to do this check when the program is actually running, known as a *runtime* check. The Pascal compiler is going to generate code something like this:

1. Figure out where in memory a is
2. If i is too high (or too low, meaning it is negative) a subscript for a, print an error and terminate the program
3. Retrieve the ith element in a

That second step implies that the compiler has squirreled away the valid index range for a, which takes a little bit of memory yet is not a big deal; the problem is that it will run the code for this check every time you make an array access, and if you are thinking, "That sounds like another one of those small runtime checks that can build up to have a nontrivial impact on the performance of your program, which the designers of C were trying hard to avoid doing automatically," you would be correct.

The C language took a different approach—one that minimizes runtime overhead. You can declare arrays as in other languages (with slightly different syntax):

```
int a[5];
```

but C does no runtime bounds checks on array access so it dispenses with that "if i is too high a subscript for a …" step entirely. If you use the expression a[i] when i is equal to 100, the C compiler generates code to calculate where in the computer's memory that element of a should be (100 spots after the first element, which has index 0) and retrieves that value. Crucially, if a in fact has fewer elements, the generated code will return whatever happens to be at that memory location, even if it is past the area that has been reserved for the a array and might belong to another variable, or be in an area of memory that isn't currently used for anything and has a random value in it, or causes a crash when accessed.

What's more, C merges the concept of arrays with the concept of *pointers*. A pointer is a variable that contains the address of another variable, which is useful for constructing certain data structures in memory; Pascal also had pointers. According to the original C book, "Pointers have been lumped with the goto statement as a marvelous way to create impossible-to-understand programs." Equating something with GOTO is no compliment, but as the book goes on to say, "This is certainly true when they are used carelessly, and it is easy to create pointers that point somewhere unexpected. With discipline, however, pointers can also be used to achieve clarity and simplicity."[16]

The insight that the designers of C had was that pointers and arrays are really the same thing: a pointer points to a location in memory, and an array also points to a location in memory, that being the address of the first element. Dennis Ritchie called this "the crucial jump in the evolutionary chain" between the language on which C was based (called BCPL) and C.[17] One benefit is that pointers can be faster than regular array lookup. Recall my breakdown of what happens on the array access a[i]:

1. Figure out where in memory a is
2. If i is too high (or too low) a subscript for a, print an error and terminate the program
3. Retrieve the ith element in a

Besides the fact that C dispenses with step 2 entirely, step 3 involves multiplying i times the size of a single element of a. Using pointers instead, iterating through an array can be done just by adding the size of each array element to a pointer as opposed to redoing the multiplication each time, and addition is faster than multiplication.[18]

Furthermore, while arrays are declared with a fixed size, which C does still support, pointers are variables and can be set to point anywhere. In C, the particularly useful place that they can be set to point is the location returned from an API that allocates memory from the system. And since pointers and arrays can be used interchangeably, you can wait until runtime to decide how large an array you need, allocate it dynamically, and assign the result of the allocation to a pointer, and then proceed to use the pointer as if it was an array. The result is not wasting memory by declaring a larger array than you need, but still being able to use the more readable array syntax.

Or you can continue to use the pointer. A few of my professors at Princeton were Bell Labs employees on short-term leave, and they would occasionally show us tricks of the trade gleaned from the halls of UNIX/C-Land. I can still recall the day that a visiting professor from Bell Labs named Henry Baird demonstrated how to change a loop from using array access to pointer arithmetic, which makes your code harder to understand but also a little faster and infinitely cooler. My attitude toward pointers could be summarized in a dialogue like this:

Wise Programmer: I have good news and bad news about pointers.
Adam: What's the good news?
Wise Programmer: They make your code faster.
Adam: Got it! Faster is cool.
Wise Programmer: They also make your code harder to read.
Adam: I see.
Wise Programmer: Faster but harder to read. That's the story on pointers.
Adam: I thought you said there was bad news?

For a final trick, C built on those three ideas—explicitly declaring different-size integers, avoiding overhead on array lookup, and allocating arrays dynamically using the array/pointer equivalency—to come up with a clever way to handle strings.

Recall that a string is a sequence of text: "hello" is a string. In machine/ assembly language, computers deal with numbers, not strings. To store strings, you first need an agreed-on way to encode your string as a series of numbers. The most common encoding system is known as ASCII (American Standard Code for Information Interchange), which in its simplest version uses 7 bits to represent various characters; the encoding of printable characters goes from 32 (which is defined to be a space) up to 126 (which is the tilde, ~). That's still enough room to fit all the uppercase letters (ranging from 65 for *A* to 90 for *Z*), lowercase letters (from 97 for *a* to 122 for *z*), and all other common punctuation symbols. The numbers 0 through 9—that is, the actual printed characters "0," "1," "2," and so on—are encoded from 48 to 57.

So my first name, "Adam," would be encoded as four numbers,

65
100

97
109

which would then be interpreted as *A*, *d*, *a*, and *m* by whatever software was dealing with them as long as it knew this was an ASCII encoding.

This all works fine, but the problem with strings is that they are not bounded the way that numbers are. A number is defined to be a certain bitness and will always use that much storage, but a string will need one byte of storage for each ASCII character. And while you can often conveniently ignore the chance of a signed 16-bit integer overflowing past 32,767, in the case of a string you need a place to store all those bytes. Concern/laziness about dealing with multiple strings, each of potentially long length, was the underlying reason behind the decision to have a single shared "street address of the last doctor we loaded from the database" string in the Dendrite code, which led to the mysterious bug I discussed in the previous chapter.

To handle this, languages such as IBM PC BASIC would keep track of the length of any string and automatically allocate memory (from the currently unused memory, which is known as the *heap*) as the length of the string changed, and then copy the string to the larger location if needed, with all this hidden from the programmer. This was fine, except that copying the string to the new location took some small amount of time. Worse, a heap allocation could fail at any time, especially on a computer with limited memory. You could write code that made your string one character longer and have the system discover, at the moment that code was running, that it was out of heap memory. Since the system couldn't perform the operation you requested, and the following statements in your code presumably depended on that string being valid, the program had no choice but to stop completely (in the case of IBM PC BASIC, with an "Out of string space" error).[19] On top of that, the longest-allowed string in IBM PC BASIC was 255 characters, which is a pretty good equivalent of the "well, it probably won't overflow" attitude of using 16 bits to store a number in that it generally works but is not guaranteed to always be enough room.[20]

Other languages had the programmer specify, when they declared a string, the maximum number of characters it would be; WATFIV and Pascal worked this way. This meant that you had to guess the longest-possible length your string might have, and waste a lot of memory if you guessed too high, and if you guessed too low and your string wound up too long, the compiler-generated code would detect this and then kill your program anyway.

The designers of C came up with a solution that neatly resolved all of this. A string was defined to be an array of 8-bit (one byte) numbers, a type known as a char, with each array element holding one encoded character (in ASCII, typically). Normally when doing any operations with arrays in C, you need to store the length of the array in a separate variable, because C internally does not track array length. This led to a lot of C functions with signatures like this:

```
myfunc(int a[], int n)
```

where a is an array (with each element being an int), and n is the length of the array; this allows myfunc() to know how many elements the array contains so it can avoid walking off the end of the array. In languages like C#, which does track the length of arrays internally and provides an API to retrieve it when needed, that second parameter is not necessary.

But since strings are used so often, a convention was adopted in C allowing a string to be passed around via a single variable, the way numbers could be. The single variable was a pointer to the first element. What about the length of the string? The convention was that the end of the string was indicated by an array element with the value of 0 (the actual value 0, not the encoding of the character "0," which is the number 48). Any code that needed to know the length of a string would scan through the array until it found the char with a value of 0 (an API called strlen() was supplied that did this for you). Incidentally, this is where the Hungarian variable-name prefix sz came from—it stands for "string, zero-terminated."

So in C, the string "Adam" was stored in an array of five char values (8-bit numbers)—which occupies 5 bytes, and is also known as a 5-byte *buffer*—containing the following 5 values:

```
65
100
97
109
0
```

This allowed strings to be allocated dynamically at any length, the same as any other array, but the entire string could be referred to with a single variable, with no extra length value needed. Furthermore, it meant that unlike languages that treated strings as a special other type and needed a specific API to perform operations such as "extract this substring from within this larger string," in C you could access individual characters in a string the same way you would access individual elements of an array. This being C, it also meant that the indexing into strings, like all array indexing, was not checked for the index being valid. And although there were helper APIs to make it easier, programmers had to write their own code to ensure that the memory buffer they allocated to hold a string was large enough for that string—and the final 0 value as well. Most famously, the length of that "Adam" string (as returned by `strlen()`) was 4 because it had 4 characters in it, but the amount of storage needed, if you wanted to make a copy of it, was 5 bytes due to the extra 0 value. Programmers writing string manipulation code in C had to be aware of this at all times.

For example, the 3-letter string "lap" takes 4 bytes of memory (the *l*, the *a*, the *p*, and the 0), and the 3-letter string "dog" also takes 4 bytes of memory (the *d*, the *o*, the *g*, and the 0), but the string "lapdog" takes up 7 bytes of memory—which as you notice, is neither 3 + 3 nor 4 + 4, which led to programmers having to add and subtract 1 from a lot of string-length calculations, leading to all sorts of "off by one" math mistakes where the extra byte for the 0 was not accounted for properly. This might turn out to be harmless, if the byte just past the end of the buffer wasn't used by any other variable, but it might also cause big problems—in particular, you might write the terminating 0 into another variable's memory that was past the end of your buffer, then have that variable replace the 0 with something else, and suddenly your string would be seen as continuing on until it happened to run into another 0 in memory.[21]

The trade-off made in string manipulation perfectly captures the heart of C: it removes any unneeded performance overhead, but also removes automatic safeguards. You had complete control of your string buffer allocations, were never required to use more memory than you needed, and did not have any runtime checks slowing down indexing. It did make reading code harder; you spent a lot of time looking at string-length-calculating and string-memory-allocating code trying to convince yourself that it was correct, typically by running through examples in your head using short strings ("Let's see, if this string is 3 bytes long, then this number will be calculated as 4, and we'll allocate that much memory here ..."). But if you did everything right, you got the functionality you needed with the performance you wanted.

And performance, not clarity or maintainability, was definitely the focus at the time I was in college. When I was a senior at Princeton, the computer science department held a coding contest in which we were asked to write a program to solve a specific problem; the only criterion (besides the program working) was how fast it ran.[22] The winner won a prize at the engineering convocation, alongside civil and aerospace engineers who presumably had worked on projects that were evaluated on something other than pure speed.

Yet to me, this all seemed perfectly normal and even encouraging. Of course performance would be the focus. What else would be? Everything that C did to clear out performance obstacles felt like taking off your tight shoes at the end of the day. I cheerfully wrote code to calculate string lengths and check array bounds, and never thought that any other way would be better. The fact that Pascal programmers were willing to trade away performance in exchange for not having to deal with these complications was seen as an indictment of their language, them, or both.

After honing my pointer-manipulation skills at Princeton, I graduated and went off to work writing C code at Dendrite. Although this was before the invention of the World Wide Web—websites and browsers—the Internet, or at least a prototypical version of it, did exist, with many machines connected together. Many of those machines were running UNIX, which as mentioned was written in C—not just the core of the operating system, but a lot of the programs that ran on it.

November 1988, a few months after I began work, brought an object lesson in the downside of the C performance trade-offs. A Cornell computer science graduate student named Robert Morris secretly released a program that came to be known as the "Morris worm." Starting from a computer at the Massachusetts Institute of Technology (MIT), the program copied itself to any other computers it could get access to, and from there connected to any other computers it could reach, in a rapidly growing circle.[23]

Although the Morris worm did not do anything malicious, such as deleting files, it would replicate itself back to the same computer over and over again, and the overhead of having the program running so many times on a single computer would degrade that computer's performance. The UNIX systems were meant to support multiple people remotely logging on to them

at once (which is why they were known, in an apropos twist, as "hosts")—
this was the setup we had in college grinding away in von Neumann, with
all of us using terminals to connect to a single UNIX machine—and users
on an infected machine would notice it becoming gradually less and less
responsive until the machine was unusable.[24] Morris had meant for it to
propagate much more slowly and therefore remain undetected for longer,
but he miscalculated how fast it would spread—a bug in his code.

After some frantic forensic work over a few days, the worm was ultimately contained,[25] and Morris was sentenced to three years' probation,
eventually winding up as a professor at MIT, and later named an ACM Fellow. But it's instructive to look at how the worm propagated itself because
it's a direct result of the choices made in designing C, in particular the
string handling.

The Morris worm took advantage of a UNIX utility called "finger," which
was used to query information about a user on a remote machine. You
could type

```
finger joe@mit.edu
```

and it would return information about the user "joe" on the machine mit.
edu, if there was such a user (the program still exists today; there is even a
version included as part of Windows).

Finger was like an early version of a personal website, except that on
the client end, you ran the finger command instead of a web browser, and
it was limited to printing out certain specific information about a user,
including the contents of two files that a user could create for this purpose,
named .plan and .project.[26]

For this to work, there had to be a program running on the remote computer waiting to receive the "tell me more about this user" message that
the finger client program would send, in the same way that a web server
waits for the initial message from a browser. This program was known as
the *finger daemon*, with the term daemon being commonly used for this sort
of thing on UNIX (it's the Latin form of a Greek word meaning "power" or
"god," which my dictionary defines as "a guardian spirit").[27] Although an
individual user could choose to not allow finger to return information to
remote people, the finger daemon itself had to listen to incoming messages
from any remote computer since it needed to determine which user the
request was for before it could decide whether to respond or not.

When two machines communicate like that, there has to be a convention, known as a *protocol*, on what information will be transferred—the code in the finger client and the finger daemon have to be in agreement. In this case, the protocol was simple: the finger client would send a username to the finger daemon, and the finger daemon would reply with several lines of information about that user, which the finger client would then display unchanged. While the message coming back from the finger daemon to the client could be of somewhat-arbitrary size, the finger daemon only expected a short message from the client—just long enough for a username.

There was nothing stopping somebody, however, from writing their own finger client that sent its own messages to the finger daemon on another computer, and there was also nothing stopping that new finger client from sending a much-longer message to the finger daemon than it expected. Of course the official finger client would not do any such thing, and part of the problem that led to the Morris worm was that nobody could imagine why anybody would write their own finger client that would misbehave in this way.

To explain why this could cause such problems, we need to review how programs typically use memory. A program that needs memory to store data will call an API that the operating system provides that hands out available heap memory on request. If a program needs a 500-byte buffer, it calls the operating system and says, "I need 500 bytes," and the operating system hands back the location of 500 bytes of memory that it has not previously handed out, and keeps track of this so it won't give that memory out again until the program calls a different API to say, "I'm done with those 500 bytes."

The "location" that the system hands back is just a number; conceptually, if a computer has 1 megabyte of memory, which is actually 1,048,576 bytes, then the available bytes of memory are identified by a number between 0 and 1,048,575. When the system hands you back 500 bytes of memory, it is saying, "Your 500-byte piece starts at [as an example] address 652,000," which means you can now use addresses 652,000 through 652,499. In C, this number 652,000 would be stored in a pointer, let's say named p, and using the magic of the pointer/array equivalency, you could now access those bytes as p[0] through p[499].

As I have discussed earlier, C doesn't check that your array index is valid, so you could access p[1000] (memory address 653,000, in this example) and be reading or writing memory that was intended to be used for something else, but hopefully your code has recorded somewhere the fact that p only points to 500 bytes of memory and won't do that.[28]

There is a part of memory that is carved off by the system for a special purpose, unavailable for general allocation. This is called the *stack* and is used for keeping track of the layers of function calls. When a program calls a function, the parameter values are stored on the stack, as is the *return address*—the place that the code should jump back to when the function completes. The return address is a memory address like any other, because the actual code that is being run—all the bytes of machine language—is also stored in memory. It's another pointer, which happens to point to bytes of code. Essentially, when the system loads a program it puts the code in one part of memory, reserves another part for the stack, and makes the rest available as the heap, for memory allocation at runtime.

Looking back on the stack, you can see the parameters to the current function and its return address, and then before that you see the parameters to the function that called that function and *its* return address, and so on; all the layers of functions, at any given point when the program is running, have their parameters and return addresses stored on the stack. It's a series of breadcrumbs for the processor to find its way back to where the program started. In addition, any local variables for a function—variables declared for temporary use inside the function—are on the stack as well.

Figure 4.1 shows what the stack would look like (with a few irrelevant items not depicted)[29] for the code on the left, where code calls function A, which then calls function B, which in turn calls function C, and we are currently running code in function C. Note that the stack grows downward in memory (that is, toward lower-numbered memory addresses), so looking back on the stack means looking up, toward higher-numbered memory addresses. To call a function, first the parameters are pushed on the stack, and then the address that the function should return back to when it is done. The calling code then jumps to the start of the function, where the first thing the function does is run code (which the compiler has automatically generated) to reserve space on the stack for its local variables.[30] In the diagram the parameters, such as A's first parameter, are described by the name they are declared as in A's parameter list—in this example, height. But realize that in the code that calls A, the parameter that becomes height inside of A is the variable cur_h in the calling code. So the calling code

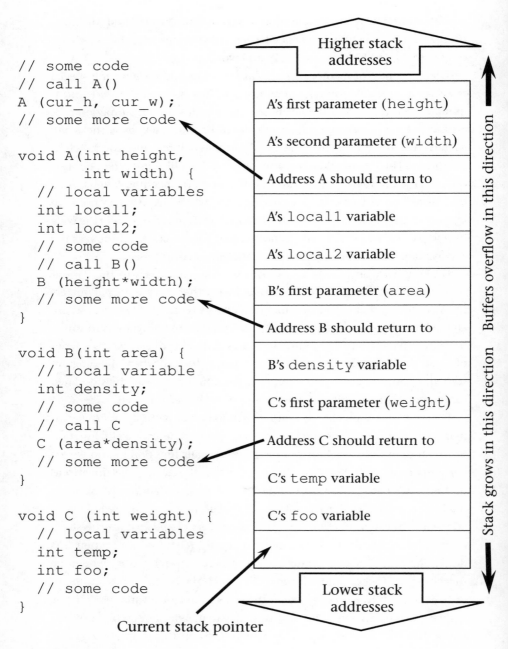

```
// some code
// call A()
A (cur_h, cur_w);
// some more code

void A(int height,
        int width) {
  // local variables
  int local1;
  int local2;
  // some code
  // call B()
  B (height*width);
  // some more code
}

void B(int area) {
  // local variable
  int density;
  // some code
  // call C
  C (area*density);
  // some more code
}

void C (int weight) {
  // local variables
  int temp;
  int foo;
  // some code
}
```

Higher stack addresses

A's first parameter (`height`)

A's second parameter (`width`)

Address A should return to

A's `local1` variable

A's `local2` variable

B's first parameter (`area`)

Address B should return to

B's `density` variable

C's first parameter (`weight`)

Address C should return to

C's `temp` variable

C's `foo` variable

Lower stack addresses

Current stack pointer

Buffers overflow in this direction

Stack grows in this direction

Figure 4.1
Code is on the left, and the stack that results is on the right (slightly simplified)

pushes cur_h on the stack, and inside of A that same value on the stack is retrieved as height.

There is a processor register called the *stack pointer* (the "current stack pointer" in figure 4.1 shows where it would be pointing when running code inside C), which tracks the next available spot on the stack.

Let's say you want to use a 500-byte buffer inside a function. There are two ways to do this in C. You can do a dynamic allocation from the heap, calling the memory allocation API malloc():

```
char* c = malloc(500);
if (c == NULL) {
    // the allocation failed, deal with this
}
```

or you can declare it as an array on the stack:

```
char c[500];
```

Due to the array/pointer equivalency in C, any code that accesses the 500 bytes inside the function will use c in an identical way no matter which of the two you use; the compiler will hide the internal differences. But in the first example, calling malloc(), the only thing on the stack is the pointer c itself—a single value that will occupy 32 or 64 bits on modern computers. In the second instance, the entire 500 bytes are on the stack. As you can see, it is less code to allocate it on the stack, and you don't have to call an API and worry about the allocation failing, which would require even more code to deal with. And it runs faster, since the space is allocated just by adjusting the stack pointer. Moreover, when any function completes and returns back to its caller, the stack pointer moves back to where it was in the caller, so you get automatic cleanup of any local variables allocated on the stack—with no need to call another API to free a stack array either. Faster and less code, as usual, are both considered to be unalloyed goods.

One important difference between dynamic allocation and allocating on the stack is that when dynamically allocating, you can specify a size that isn't known until runtime; that "500" parameter to malloc() could instead be any variable. Meanwhile, the local stack allocation as shown above requires a constant value (certain versions of C allow you to allocate on the stack with a variable size, but it's not standardized).

So if you are concatenating strings together into a new string, the correct thing to do is to determine the length of the original strings (the first two lines below) and allocate the right amount of memory (the third line, with the + 2 needed to account for the space and the terminating 0 character) before calling the APIs to actually do the concatenation (the last three lines: first copy over the first name, then concatenate on the space and the last name):

```
int firstnamelen = strlen(firstname);
int lastnamelen = strlen(lastname);
wholename = malloc(firstnamelen + lastnamelen + 2);
if (wholename == NULL) {
     // the allocation failed, deal with this
}
strcpy(wholename, firstname);
strcat(wholename, " ");
strcat(wholename, lastname);
```

But it is oh so tempting to simply ask, "How long can a name possibly be," and then pick a number such as 256:

```
char wholename[256];
strcpy(wholename, firstname);
strcat(wholename, " ");
strcat(wholename, lastname);
```

This is less code to write, and as a bonus it runs slightly faster and you don't have to deal with the "allocation failed" part, which could get ugly (and you also have to free your heap allocation at some point). Note that the last three lines of code, the ones that produce the concatenated name, are the same; by storing the string on the stack, you are saving 3 API calls plus having to deal with an error case. Isn't that convenient? And the only risk is that somebody has a first name/last name combo that is longer than 256 characters (254, actually, since you need room for the space and final 0)—and how likely is that, really?

Now I can explain the finger daemon bug that the Morris worm exploited. When reading the request from the client, which was only expected to contain a username, the finger daemon code did the quick "let's just assume the request buffer won't be bigger than a certain size and allocate room for it on the stack" trick, and the certain size it picked was 512 bytes. And the API it was calling, which was called gets(), took as its single parameter a string buffer to read the data into, but with no way of indicating how long the buffer was; it assumed the calling code knew what it was doing and the buffer was long enough. Specifically, gets() was designed to read "a line of text"; recalling that the encoding of printable characters in ASCII goes from 32 to 126, gets() reads data until it sees a byte containing the value 10—a nonprintable character in the ASCII encoding known as the line feed (LF). The string buffer that gets() was passed was a pointer to a location in memory; gets() would read in the first byte, store that at the pointer location, advance the pointer by 1 byte, and loop back to read the next byte. It would continue doing this until it saw the LF character, completely unaware of what size had actually been specified when the string buffer was allocated. It was entirely possible that gets() could go past the end of the buffer—a situation known as a *buffer overflow*.

The design of gets() is terrible; honestly, I have no idea what they were thinking. In the string concatenation example above, you at least have the opportunity, if you choose to take advantage of it (as the first version does), to do the actual calculation of how much storage you need, and even have a decent chance of getting it right (just remember to add 1 to account for the final 0 byte!). But in this case, no matter how big a buffer you passed in, gets() could still overflow it if it didn't encounter an LF character in time.[31]

This is another example of being stuck with the implementation of an API you are calling; gets() is the gantlet that the program must run to get the data it needs. That API will continue to read data until it sees that LF character, and you have no control over that; no matter how big a buffer you pass in, you don't know if gets() will read more than you allocated because you don't know what actual data it will be reading at runtime. Inside of gets(), meanwhile, the code doesn't know how long the buffer you passed in was allocated to be, nor does it know if it was on the stack or heap. The only saving grace for gets() is that it is somewhat obvious that it can't know how long the buffer is, because there is no parameter specifying the length. So hopefully people will realize that it is hopeless and not call it.

Except that the finger daemon did call it. Somewhat depressingly, the finger daemon could easily have called a more general API that reads from any file and told it to read from a special file called the *standard input*, which is what gets() does, and that API does take a parameter indicating the maximum length of the buffer:

```
fgets(buffer, 512, stdin);    // instead of gets(buffer)
```

So what are we talking about—typing 9 more characters (11 if you count the commas, 13 if you count the spaces) to switch to an API that isn't at risk of a buffer overflow and prevents the Morris worm?[32] Why did the author of the finger daemon not do that? For that matter, why was gets() designed in such a horribly broken way, when it could have mimicked fgets() and taken the extra length parameter? These are historical questions whose answers are unknown to me; as a small consolation, gets() was eventually excommunicated from the C language standard so it can no longer infect future code (although I'm sure there are a lot of old systems out there running code that has calls to gets() buried inside).

In some cases, when a buffer is too short to hold the data, the program doesn't have an obvious recourse. It's like filling in a form by hand that has a square for every letter and discovering that the designers of the form assumed that no last name would be longer than 15 characters, and you have a 16-character last name. What do you do? There's no right answer. But here the answer is simple. Since a 513-plus-byte finger request is clearly wrong, you can ignore it. In fact, if you are going to switch to using fgets() and ignore overlong requests, while you are at it you can use a much smaller stack allocation, such as 64 bytes.[33] The middle ground of 512 is much longer than you will ever need for a valid request, but it isn't long enough to protect you against an invalid request, so what is the point?

The fact that the finger daemon buffer that overflowed was on the stack was the worst part of it. When running inside a function, as shown in the earlier diagram, the stack is arranged with the local variables immediately below the parameters and return address; the stack overall grows downward, but access to a buffer goes upward. Therefore, the parameters and return address were just past the end of the 512-byte buffer, which was declared as a local variable on the stack rather than allocated dynamically from the heap because that takes less typing and stack allocations magically never fail (in fact the stack is just memory and can run out of room, crashing the program, but programmers ignore this).

A buffer overflow is always bad, because you will likely corrupt whatever memory is just after your buffer, but at least if the buffer is somewhere in the heap, the likely outcome will just be strange behavior and/or making the finger daemon crash, which is annoying and disruptive yet not an existential threat to the proto-Internet (although there are exceptions: the remote attack against Windows known as EternalBlue, which was the basis of the WannaCry ransomware attack in 2017, took advantage of an overflow in a heap buffer; it relied on clever tactics regarding exactly what messages it sent to the computer to ensure that the heap buffer just after the one that was overflowed could be usefully and reliably modified).[34] When you have the buffer on the stack, your malicious finger request is going to stomp on the most dangerous piece of data in memory: the return address of the function that is calling `gets()`. This is data that the processor is going to blindly obey when it decides where it needs to jump back to when this function is finished running.

By another appalling lack of serendipity that I won't get into, you can set the return address to point into an earlier part of the buffer itself, and the computer will happily jump to any return address, even one that is in the area reserved for the stack versus where code is normally loaded. So the malicious finger request (such messages are generally known as *exploits*) can send over the actual code that it wants to run (known as the *payload*) in the same bogus finger message that it uses to overflow the stack buffer.

The payload has to be in machine language, since that is what the processor expects to see in memory when it is running a program. But you have on the order of 512 bytes to craft your payload—more than enough room to hold the fairly simple exploit code that the worm used (the one gotcha is that your machine language code can't have a byte with the value of 10, since that would be interpreted by `gets()` as an LF character and stop it reading data).

If you had been intentionally designing a system to allow remote exploits, you could hardly have done a better job. This is the "worst performance-focused feature of C" that I mentioned at the beginning of the chapter. The lack of safeguards on arrays, ease of allocating local buffers, and arrangement of the stack, as well as the appeal of UNIX as bait, make C the perfect Manchurian sleeper language. To quote Arwen in *The Lord of the Rings*, "Not till now have I understood the tale of your people and their fall. As wicked fools I scorned them, but I pity them at last. If this is indeed, as the Eldar say, the gift of the One to Men, it is bitter to receive."[35]

It's not that the concept of malicious code was unknown in 1988. The book *Computer Viruses* had appeared the year before, talking about the dangers of unknown code running on your computer and discussing the various means of infection.[36] But all the viruses explored in that book were spread due to what could be loosely called "human error." Although the usage is not precise, the term *virus* generally means an exploit that has to be spread "manually," such as by a user copying a file, as opposed to a worm, which can spread with no user action. The attacks up to this point had been viruses, not worms.

For example, there was a previous "infects the Internet" virus in late 1987 called Christmas Tree, which was spread via e-mail, and when run, e-mailed itself to everybody else it could find.[37] But Christmas Tree depended on the user receiving an e-mail that essentially said "run this program, no questions asked" (before infecting your friends, it printed out a Christmas tree made of asterisks for your enjoyment). And there were other viruses that depended on a trusting user running an infected program, which would then infect other programs it found on your computer (this is why e-mail programs today will refuse to open, or in certain cases even transmit, types of file attachments known to be risky).

All these could be dismissed by programmers as not their fault: "Well, if the user does something stupid, then they have to face the consequences," where the stupid action was running a program whose contents you were ignorant of (in the case of Christmas Tree, which was an interpreted script in the language REXX, just scanning the code would make it obvious it was up to something devious; of course this would require you to understand REXX, but I suppose not being a programmer was also lumped into the category of stupid user behavior).[38] At one unfortunate point in the history of computers, when most users were also programmers, programmers would refer to the rare nonprogramming users as *lusers*—a play on the word *loser*. Blindly running a virus script was something that a luser would do.

The Morris worm was different: it infected machines with no action by the user. The cardinal sin was a onetime mistake by a programmer, and from that point on there was nothing that could have been done to prevent it, until the worm was out there and wreaking havoc, and people were inspired to deploy a new version of the finger daemon that had the buffer overflow problem fixed. There was even a suggestion at the time that some people knew the code in the finger daemon was risky but ignored the problem, presumably because of a failure of imagination; they knew the real finger client would never send a message anywhere near 512 bytes, and it never occurred to them that somebody would write a program to maliciously generate invalid requests.[39]

This could have functioned as a wake-up call to C enthusiasts, but it did not. Certainly there was no outcry against C-style buffer manipulation, although there was movement to excise gets() specifically from people's code. The Cornell Commission, in its Summary of Findings, after first (correctly) laying the blame with Morris, goes on to state,

The fact that UNIX, in particular Berkeley UNIX, has many security flaws has been generally well known, as indeed are the potential dangers of viruses and worms in general. Although such security flaws may not be known to the public at large, their existence is accepted by those who make use of UNIX. It is no act of genius or heroism to exploit such weakness. A community of scholars should not have to build walls as high as the sky to protect a reasonable expectation of privacy, particularly when such walls will equally impede the free flow of information. Besides, attempting to build such walls is likely to be futile in a community of individuals possessed of all the knowledge and skills required to scale the highest barriers.[40]

I can accept their view of the Internet as a "community of scholars" rather than the early version of the most important thing ever, since this was 1989, but beyond that it's hard to know where to start with this. The casual acceptance of exploitable code? The backhanded denigrating of Morris's technical skills? The conflating of "don't call the verkakte gets() API" with impeding the free flow of information? The main thing is that this accurately captures the palms-to-the-sky, what-are-you-gonna-do attitude of programmers toward buffer overflows.

It is also hard to separate C from UNIX; they were made for each other. Niklaus Wirth described it this way: "In its tow UNIX carried the language C, which had been explicitly designed to support the development of UNIX. Evidently, it was therefore at least attractive, if not even mandatory to use C for the development of applications running under UNIX, which acted like a Trojan horse for C."[41] The finger daemon required interacting with the network, which is not something that any programming language supported with its standard API, but on UNIX, where C was the language of choice, there were API available in C that let you send and receive network messages, so naturally the finger daemon was written in C.

"From the point of view of software engineering, the rapid spread of C represented a *great leap backward*," continued Wirth,

It revealed that the community at large had hardly grasped the true meaning of the term "high-level language" which became an ill-understood buzzword. ... The widespread run on C undercut the attempt to raise the level of software engineering, because C offers abstractions which it does not in fact support: Arrays remain without index checking, data types without consistency check, pointers are merely addresses where addition and subtraction are applicable. One might have classified C as being somewhere between misleading and even dangerous. But on the contrary, people at large, particularly in academia, found it intriguing and "better than assembly code," because it featured some syntax.[42]

I wasn't just intrigued by C because it "featured some syntax"; I found the very things that Wirth is criticizing to be the most appealing features of C. Indeed, when I heard about the Morris worm exploiting a buffer overflow in C code, I thought to myself, "That's kind of clever!" Even one of the reports on the worm referred to the finger daemon buffer overflow as the "neatest hack" that the worm did.[43]

More important, there was no feeling after the fact that anything was fundamentally wrong with writing network-facing system tools in C. The episode was viewed as the result of a single mistake, which had been fixed in newer versions of the finger daemon, and the problem was solved. Nobody thought, "Gee, there might be a lot more of these out there." As mentioned, programmers were self-taught, and this tended to lead to a certain lack of introspection. It certainly never occurred to me that my code might have similar bugs. I was confident that I could get the math right when dealing with buffer and array lengths.

Whatever rules you feel that the "structured programming" movement was trying to enforce, it clearly had an effect on what programmers did. Even if it boiled down to "don't use GOTOs indiscriminately," the message managed to get through to me, despite having taught myself to program in GOTO-infested Fortran and BASIC. But C was like one of those foreign spies raised in a simulated US village who passes the eye test when they arrive in the United States: it clearly was structured, in that it supported the sequence/selection/iteration model and did not require GOTO statements. Yet a dark heart lurked inside it, which was not only missed by people like me but also celebrated as representing freedom from our old fetters.

The C language didn't only enable charismatic megafailures such as the Morris worm. Code that accidentally indexed off the end of an array could cause all sorts of subtle failures. If your code read a value past the end of an array, the program might crash, but it also might successfully read memory intended for another use, which might vary each time you ran the program. If your code *wrote* to memory past the end of an array, you might corrupt another data structure in memory, again in an unpredictable way; when you tried to debug the problem, you would have no way of knowing what code had written the bad value. You might have a bug that happened every time, but in a slightly different way, which is extremely tricky to diagnose. At one point during the development of Windows NT, a fellow programmer was debugging such an error, where his data structure was being randomly corrupted by unknown code. I can still recall the bleak look on his face as he contemplated scanning reams of code looking for the problem.

There were naysayers at the time C appeared. The above quote from Wirth is from a retrospective article he wrote in 2008, but he was always opposed to the amount of rope that C handed to programmers. In 1979, Wirth came out with the language Modula-2, a successor to Pascal, which was also designed to be powerful and performative enough to write an operating system in (specifically, a system called Lilith that Wirth also wrote, so I assume the performance claim was accurate), and that had bounds-checked array lookup. Modula-2 was swept aside in the C/UNIX landslide; the tagline "an improved version of Pascal," if it ever reached my ears, would undoubtedly have failed to impress, and Wirth, a true titan in the annals of computer science, would have been unfairly dismissed as a grumpy guy who probably couldn't write correct pointer code.

OK, fine, you are saying. Buffer overflows can happen. Yet are programmers really that stupid? Maybe you can grant them that people make arithmetic mistakes counting bytes, and perhaps you can understand the part about trusting the return address on the stack, if you don't think there is any way for it to be tampered with. But why, for heaven's sake, are you even storing the untrusted user's message anywhere near the trusted return address? Why not put the user's message somewhere else, where even if you botched the handling of it, it wouldn't allow you to be taken over by an exploit? And why not write the whole thing in a language that keeps track of buffer lengths and prevents them from overflowing?

This last is an excellent question. You can do it that way, and a lot of newer languages do, which is why the C# string manipulation code that we have seen is so much simpler than what is required in C. But the question boils down to a deeper question in programming—one known as "errors versus exceptions." And therein, as they say, lies a tale, which I will get to a few chapters from now. First, though, I'll consider the more prosaic question of how programmers determine if their software is going to work at all.

5   Making It Right

The term "software crisis" is not heard as much today, but there was a lot of talk about it in the 1960s. As Matti Tedre relates in his well-researched 2015 history *The Science of Computing: Shaping a Discipline*,

In the course of the 1960s, computing's development curve was about to break. The complexity of computer systems had all but met the limits of the popular software development methods of the time. The crisis rhetoric entered computing parlance over the first half of the 1960s. ... By the end of the 1960s project managers, programmers, and many academics alike had grown so weary of the blame and shame that immediate improvements were deemed necessary.

(Tedre also comments, "The crisis talk that was rooted in the 1960s and popularized in the early 1970s has remained with computing ever since— whether or not a decades-long quagmire of problems should be called 'crisis' anymore.")[1]

In 1968, the North Atlantic Treaty Organization (NATO) sponsored a conference in Garmisch, Germany, bringing together academia and industry to discuss these problems. This conference was a seminal moment in the history of software engineering (for one thing, it popularized the term), although its net effect is unclear. The conference report laid out all the problems in software that continue to plague us today—reliability, management, scheduling, testing, and so on.[2] Then again, many writers were describing the same problems at the same time; the issues weren't a secret to anybody who had written nontrivial software. Some have described the conference as a critical point in the schism between industry and academia, especially after a follow-up conference in Rome the next year devolved into a more explicit division between theorists and practitioners.[3] Then there is a theory that the whole thing was part of a fiendish plot by Dijkstra to get people to pay more attention to his notions of structured programming.[4]

Whatever the NATO conferences did or did not accomplish, the root cause of all this concern was that programmers couldn't figure out how to write software without bugs. The problems in software engineering to this day either relate directly to bugs (reliability and testing) or to dealing with the unpredictability of bugs (management and scheduling).

Just what are software bugs? Although it might feel like a bug when software doesn't perform the way the user expects, from a programmer's perspective a bug is when the software doesn't perform the way the *programmer* expected. To them, bugs are surprises. There are books devoted to the issue of what programmers expect versus what normal people do, and who should be designing the software and setting the expectations, but that is a separate topic.[5] Certainly, many changes to software are made because the user wanted the software to operate differently from the way it was designed to work, but I consider those to be "enhancements" rather than "bugs." Donald Knuth, while categorizing the changes he made to a large software package, provided the cleanest differentiation: "I felt guilty when fixing the bugs, but I felt virtuous when making the enhancements."[6]

Actually, the term *bug* is used too broadly. People who study software errors talk about three levels: *defect, fault,* and *failure.*[7] Those terms are not used consistently (even by me, I'm sure); typically for software engineering, there has not been enough study of the matter to bring about any standardization. Some writers use *infection* instead of *fault,* and the word *bug* is sprinkled around to mean any or all of them. For our purposes, I'll define them as follows. A defect is an actual flaw in the code: the mistake that a programmer makes. The running program logic, viewed at a certain level, consists of manipulating data in the memory of a computer (what Dijkstra, in his complaint about GOTOs, called the *process*); at some point a piece of memory will have the wrong value in it, due to the defect in the code—this is the fault. Finally, this fault will cause an error that is noticeable to the user—the failure.

Not every code defect will cause a fault; it must be executed under the necessary conditions. If you recall the Year 2000, or Y2K, bug, one of the fears was that software that stored the year in two digits as opposed to four would make wrong decisions about durations of time when the year rolled over from 1999 to 2000. For example, the code controlling a nuclear reactor might contain this (where the two vertical bars, | |, mean "or"):

```
years_since_service = this_year - year_of_last_service;
if (years_since_service == 0 ||
            years_since_service == 1) {
    // we are OK, was serviced this year or last
} else {
    // haven't been serviced in 2+ years, help!
    initiate_emergency_shutdown();
}
```

If the variables holding the year values (this_year and year_of_last_ service) are storing only the last two digits, then this code has a defect, but it only manifests itself as a fault in the year 2000, when this_year is 00 and year_of_last_service is 99.[8] The code will calculate that it has been −99 years since the last service, which is not equal to the values of 0 or 1 that the IF statement is looking for, and decide to initiate an emergency shutdown. That value of −99 in years_since_service is the fault, and the call to initiate_emergency_shutdown() would (presumably) cause a visible failure, but the defect could lurk for years before the fault and then the failure actually arose.

Conversely, there may be no defect; storing dates as two digits doesn't require that the code behave this way. It could be written such that when it sees an unusual value like −99 for years_since_service, it assumes that we have crossed a century boundary and handles the situation correctly. This is why there was such debate, ahead of time, about what failures would result from the Y2K bug—unlike in civil engineering, say, where you can reasonably assess the condition of a bridge and how much it will cost to repair it, it is very difficult to determine how failure-prone a piece of software is. In the end the year 2000 brought some minor problems, with automated ticket machines not working or websites displaying an incorrect date, but no major disruptions.[9] Given that a lot of software had been patched or replaced in the years leading up to 2000, it will likely never be known how severe the problem would have been had it been ignored.

The code above is a dramatization of a bug; most are more subtle, and don't involve nuclear reactors and APIs named initiate_emergency_ shutdown(). And to be fair to the authors of Y2K-suspect code, many assumed that their software would be replaced long before the year 2000 rolled around, so they didn't bother writing it in a Y2K-safe way. The key point to appreciate is that when reading the code above, it is not obvious that there is a defect that will cause an API to be called when it is not supposed to be (and in many cases, the source code for old software isn't available to be read).

Not every defect will cause a fault, but every fault is caused by a defect (leaving out hardware problems, which I will do). Faults don't just happen with nobody being to blame; they happen due to defects in the code, and if the code had no defects, they wouldn't happen. Furthermore, they happen deterministically if the right repro steps are followed. Every time you execute the defective code under the right conditions—in the example above, in the year 2000, when the reactor was last serviced in 1999—the fault and failure will happen.

So you have your fault—the bad data in memory. Yet just as not every defect causes a fault, not every fault will cause a failure. To turn a fault into a failure, the bad value in memory is read by code further along, which may lead to further bad data, which is read by other code, and so on, until the effect of the bad memory becomes visible to the user—a nuclear reactor shuts down, perhaps, but more commonly an icon on the screen is in the wrong place, a character is not visible, the spreadsheet shows the wrong value, or if things are wrong in exactly the right way, the program will crash, which we'll get to in a moment. It's also possible, though, that this whole chain of events may not link up; the incorrect value in memory may not be used by any future code, so it may not matter that it was incorrect.

Consider an underage person who goes into a bar and shows identification to prove that they are twenty-one years old, the legal drinking age in the United States. Imagine that the bouncer only considers the year in which somebody was born and therefore will think a person is twenty-one even if they will turn twenty-one later this year, but have not yet had their birthday. That flaw in the bouncer's thinking is the defect (assume the bouncer always makes this mistake in a consistent way, as a computer program would). Now imagine further that the bouncer looks at the twenty-year-old's identification and decides that the person is of legal age. That is now the fault; in software terms, you could think of this as the bouncer having the "are they of legal age?" variable set to "true" when it should have been "false." Note that the defect didn't have to cause this fault—the person might have already had their birthday this year, or maybe they were eighteen or twenty-three, so the defective calculation didn't matter—but in the exact set of circumstances, the bouncer will make the mistake.

So now we have the fault, whereby the bouncer thinks that the person is of legal drinking age. This will not automatically cause the failure of letting the person into the bar. Perhaps the bouncer will decide that the person is not dressed appropriately and deny them entrance for that reason. Perhaps there will be a second identification check by somebody else who does the math correctly. Maybe the person, as they are about to enter the bar, will receive a text message inviting them to go somewhere else. Still, if none of that happens, then we will have an underage person in the bar, which is the visible failure, which could be traced back to the fault, which could then be traced back to the defect. And if you fixed the defect, by either training the bouncer to calculate ages properly or hiring a new bouncer, then that specific way of generating the fault would be removed and thus would never cause a failure.

This fix would not prevent underage people from ever being admitted to a bar; it would just mean that this particular sequence of defect to fault to failure has been prevented. You may have other faults that lead to the exact same failure, which can make it hard to figure out what is going on. You might fire that bouncer, offer math classes to the rest, and yet next week the police come by and find another twenty-year-old ordering drinks.

The repro steps of software bugs are typically reported in terms of the failure (the problem visible to the user), and debugging them consists of two parts: finding the fault (the bad value in the computer's memory), and using that to track down the defect (the code error). Finding the fault can be tricky, because what you need to find is the first fault, which may quickly metastasize into a set of faults, as variables are assigned values calculated from the faulty values of other variables. Debugging often consists of running the program for a bit, examining the contents of memory, determining them to be fine, running it a bit more, realizing that the contents of memory have gotten messed up, and then repeating this in smaller increments, trying to narrow down the point of first fault (this is why having reliable repro steps on a bug is so important). Once the first fault is found, the code defect is generally obvious, although this assumes you are familiar with the code— hence the added difficulty of debugging somebody else's code. The proper way to fix the defect, of course, is subject to the usual debate.

Broadly speaking, there are three types of failures: crashes, hangs, and just plain misbehaviors. Crashes are the most dramatic kind of failure, when the program stops running unexpectedly; they usually happen because the program tries to access memory that has not been allocated to it. As we've discussed, in C a pointer is just a number, which is not guaranteed to be the address of a valid memory location. For one thing pointers are frequently initialized to 0, known as the *null pointer*, which will cause a crash if it's used to access memory (although the first byte of memory in a computer is nominally at address 0, that location is marked as off-limits to programs). Since C doesn't do array bounds checks, any slightly out-of-bounds array index may result in a bad pointer crash (which arguably, is better than silently reading bad data, which might also happen). In languages that do have runtime bounds checking, the program will be intentionally crashed if an out-of-bounds array index is detected, which is a slightly cleaner experience conceptually, but isn't much better from the user's perspective.[10]

If a program does not crash but instead hangs, it is likely stuck in a loop. I have shown examples of simple loops:

```
FOR I = 1 TO 10
    PRINT I
NEXT
```

so you might wonder how such code can get stuck, but many loops are much more complicated.

One form of loop is called the WHILE loop, which terminates when a logical expression is false. A widely reported bug in a WHILE loop caused Microsoft's Zune music player to hang when it tried to boot on December 31, 2008. The clock on the device stored the date as the number of days since January 1, 1980, which took up less space than storing a full date and made calculating date ranges easier; in order to convert that date to a year for displaying to the user, it had this code, which starts with the year 1980 and lops off one year's worth of days from the date, until it gets to the current year:[11]

```
year = 1980;
while (days > 365) {
    if (IsLeapYear(year)) {
        if (days > 366) {
            days -= 366;
            year += 1;
        }
    }
```

```
   else {
      days -= 365;
      year += 1;
   }
}
```

If you want to read this code, it's best to first pretend that leap years don't exist, so IsLeapYear(year) is always false; then the code is starting with a days value and going through the ELSE block of the if (IsLeapYear(year)) statement, doing this over and over:

```
while (days > 365) {
   days -= 365;
   year += 1;
}
```

which is a reasonable (if slightly inefficient in this case) way to figure it out. If the number of days still to be accounted for, as stored in the variable named days, is more than 365, then add a year and subtract off those 365 days, then loop back and check again.

From there, you can see that the "true" branch of the IF covers the special case of leap years, subtracting off 366 (rather than 365) from days to account for them. Again, there is nothing logically wrong with this.

The code works correctly on any day that is *not* December 31 of a leap year. Unfortunately, on December 31 of a leap year it loops forever: IsLeapYear(year) is true when year reaches the current year, and once the code is done chopping days down to account for every year between 1980 and last year, then days will be 366 (December 31 being the 366th day of a leap year), but the code only checks if days is greater than 366; if days is exactly 366, then the code won't change it at all, and the WHILE loop will iterate again and again—it's the "Lather. Rinse. Repeat." instructions from the shampoo bottle, writ in code (meanwhile, if days is less than 366, on any earlier day in a leap year, the while (days > 365) loop will have exited already). In addition to checking for days being greater than 366, as it does, the code needs an additional check for days being equal to 366 so it can break out of the loop in that case (there are also myriad other ways to rework the code to avoid the bug).

The defect is this missing code; the fault is days remaining at 366 forever; and the failure is the visible hang—the Zune would not boot on that day. As with the Y2K bug, this defect lurked for a while before causing a fault, from the time the Zune was released in November 2006 until the first time the calendar had a 366th day in a year, on December 31, 2008 (when days entered the above loop with a value of exactly 10,593).

Crashes and hangs are rarer than the "just plain misbehaving" category. Whether code bugs cause the application to crash, hang, or misbehave is usually a matter of luck, unrelated to the difficulty of the code being written. In the Zune example, a slight tweak would produce code that didn't hang but instead reported December 31 of a leap day as "January 0," and if the code then tried to look up day number 0 in an array, it could easily crash on a bad array access. All these would appear as roughly the same code; there is nothing obvious in a defect that indicates what sort of failure the fault will cause.

Complicating the situation is that the problem may not even be in code that you can look at; as in other areas of programming, you are at the mercy of the API you are calling. You can call an API that normally works, and then suddenly it crashes, or hangs, or does the wrong thing (the Zune hang bug was not in Microsoft's code but instead in the implementation of an API that it got from another company, which Microsoft's code called during the Zune start-up sequence). Doing the wrong thing often means that the API returns an error rather than succeeding, at which point the code calling the API has to decide what to do. Does it shrug its shoulders and show the user the unexpected error? Does it call the API again in hopes that it succeeds? Does it do something clever to preserve the user's work? All this requires that more code be written, so it's up to the programmer's judgment about how likely an API is to fail, and how much work it is worth doing to deal with that case in a clean way.

One notorious example of software being at the mercy of an API it was calling was in the original version of DOS that shipped with the IBM PC in 1981. If a program called an API that DOS provided to save a file to a disk—which back then was a removable 5 ¼-inch floppy disk—and it failed, DOS itself would prompt with a message asking the user to "Abort, Retry, Ignore." Abort crashed the program immediately, and Ignore pretended that the operation worked even though it hadn't—neither of which was a particularly useful choice (as the DOS manual noted in regard to Ignore, "This response is not recommended because data is lost when you use it").[12] Retry would try again, which could handle the most trivial case (when the user forgot to insert a floppy disk in the drive), but then you would be stuck in the Retry loop until you inserted a floppy disk. From the perspective of the program calling the API, this was all done under the covers; the API call would not return until the operation had been successfully retried or the error ignored. Programs could avoid getting stuck in this DOS error prompt, but it required writing extra code, which some programmers neglected to do.[13] At one point in the early days of the IBM PC, whether a program such as a word processor could handle a missing floppy without crashing and losing data was a point of evaluation in magazine reviews. A fourth option, "Fail," was eventually added to the choices in an update of DOS, allowing the API to return to the program with an error and giving the program the opportunity to decide how to proceed.

Conceptually the same thing can happen in a car crash: the car is calling the tire "API" and asking it to provide adhesion to the road, and if the tire fails to do so, the car may crash. But the tire has been tested and certified to operate a certain way. Think of all the other engineered objects with which you interact in your daily life, such as every railing you lean on, every electric device you plug in, or every medication you take. You expect these to work every single time, over and over, without randomly "crashing," and because you expect this, the people who engineer these products put a lot of research and design effort into making sure that they do work. The tire that blows out and causes your car to crash was made by people who expect it to never fail catastrophically if used as intended, and were subject to regulations designed to prevent such failures. And most important, if it fails, they understand that they have failed in some small way and try to prevent it in the future—despite the fact that tires suffer physical wear and tear that makes failures more likely, unlike software.

When you call an API to figure out what year it is, there is no way to know if it is going to suddenly fail, hang, or crash just because we happen to be on December 31 of a leap year. As with the term *worm* from the Morris worm that I discussed in the last chapter, talk of viruses and crashes makes them sound like bad luck that can happen to anybody, and thus may be avoided due to environmental factors, but that's misleading. When people say, "It's inevitable that a large program will have bugs," they don't mean inevitable in the sense, "It's inevitable that cars will have accidents." What they mean is, "We don't have the proper software engineering techniques to root out all defects so we're not even going to attempt to remove them all—and we're not going to improve the techniques either."

Yet a lot can be achieved by aiming for perfection, even if you don't reach it. The car company Volvo has set the following goal: "By 2020 no one should be killed or seriously injured in a new Volvo car."[14] Will it achieve this? Probably not. But it sure does help focus Volvo on safety. This institutionalized acceptance of shoddiness is one of the most shameful aspects of software engineering. Software bugs are *not* inevitable, but trying to write software that never crashes is a nongoal, as they say, for the current crop of programmers. And the implicit comparison to automobile crashes silently enables this attitude.

The early days of software did feature a significant emphasis on how to produce completely bug-free software. In the years following the 1968 and 1969 NATO software engineering conferences, two books appeared on the topic of writing better software; not surprisingly, both were titled *Structured Programming*. The first came out in 1972, and was written by Ole-Johan Dahl, Edsger Dijkstra, and C. A. R. Hoare. The second came out in 1979, and was written by Richard Linger, Harlan Mills, and Bernard Witt.[15] The first book was written by leading academic computer scientists (Dijkstra attended both NATO conferences, and Hoare was at the second), and IBM veterans (none of whom were at either conference, although IBM was well represented at both) wrote the second.[16]

The first book is three long essays: "Notes on Structured Programming" by Dijkstra, "Notes on Data Structuring" by Hoare, and "Hierarchical Program Structures" by Dahl and Hoare. They lay out the basics of structured programming, as we have seen earlier, as well as talking about a few common data structures and how to break a large problem into smaller ones. This is not to criticize the work; at the time, the "structured programming" debate, which could in hindsight be summarized as the "don't use GOTOs" debate, was still being fought, so this is a worthwhile accomplishment. As Knuth enthused, "A revolution is taking place in the way we write programs and teach programming. ... It is impossible to read the recent book *Structured Programming* without having it change your life."[17]

The problem with the book is summarized in a quote from the book itself, in a section of Dijkstra's essay titled "On Our Inability to Do Much":

What I am really concerned about is the composition of large programs, the text of which may be, say, of the same size as the whole text of this chapter. Also I have to include examples to illustrate the various techniques. For practical reasons, the demonstration programs must be small, many times smaller than the "life-size programs" I have in mind. My basic problem is that precisely this difference in scale is one of the major sources of our difficulties in programming!

It would be very nice if I could illustrate the various techniques with small demonstration programs and could conclude with "... and when faced with a program a thousand times as large, you compose it in the same way." This common educational device, however, would be self-defeating as one of my central themes will be that any two things that differ in some respect by a factor of already a hundred or more, are utterly incomparable.[18]

Dijkstra's solution to this problem is to have his essay contain little code and focus more on theoretical insights, and the essays on data structuring and program structure also confine themselves to theory and small fragments of code.

The IBM crew, meanwhile, takes a noble stand with its goals, right from the first paragraph in the book:

There is an old myth about programming today, and there is a new reality. The old myth is that programming must be an error prone, cut-and-try process of frustration and anxiety. The new reality is that you can learn to consistently design and write programs that are correct from the beginning and that prove to be error free in their testing and subsequent use. ...

Your programs should ordinarily execute properly the first time you try them, and from then on. If you are a professional programmer, errors in program logic should be extremely rare, because you can prevent them from entering your programs by positive action on your part. Programs do not acquire bugs as people do germs—just by being around other buggy programs. They acquire bugs only from their authors.[19]

What about Dijkstra's point about scaling, keeping in mind that this book was written by people who had been involved in writing some of the largest software of the day and so leaving them acutely aware of the problem? The IBM trio's answer is,

It will be difficult (but not impossible) to achieve no first error in a thousand-line program. But, with theory and discipline, it will not be difficult to achieve no first error in a fifty-line program nine times in ten. The methods of structured programming will permit you to write that thousand-line program in twenty steps of fifty lines each, not as separate subprograms, but as a continuously expanding and executing partial program. If eighteen of those twenty steps have no first error, and the other two are readily corrected, you can have very high confidence in the resulting thousand-line program.[20]

Worthy goals indeed, but a thousand-line program is pretty short and still within the capacity of one person to produce. The rest of the book is about how to prove your software is correct using a mathematical approach. Still, this is the distilled wisdom of people from IBM. What did I think of this advice, when I went off to college five years later?

I thought nothing of it, naturally, since I wasn't exposed to any of it. Whether you believe in mathematical proofs or not, I wasn't taught how to know whether the software I wrote worked. But looking back, it's clear that the time for mathematical proofs was passing. For simple code like a sort routine, you can prove that your algorithm is correct: the array begins in this state, the loop iterates this many times, after each iteration one more element is sorted, and therefore the array is sorted at the end. This is a standard mathematical technique known as induction. The problem is that modern software is vastly more complicated than that. A failure today along the lines of "my word processor is showing this character in the wrong place" involves extremely complicated code that has to factor in the margins of the document, the font being used, whether the character is super- or subscripted, how line justification is being done, the size of the application window, and a host of other factors; the code winds up being a tangle of IF statements and nonobvious calculations.

Think back to the Zune leap year bug: the defect was not in the algorithm but rather in the implementation. Many annoying bugs, when finally excavated, turn out to be nothing more than a typing mistake by the programmer. The IBMers' *Structured Programming* book was a curious atavism—a book advocating techniques that were only effective for short programs, produced by people who had worked on large programs.

If there was a guiding spirit of software quality that attended my college years, it came not from the academics or IBM veterans but instead from a third fount of programming wisdom: Bell Labs, the source of UNIX and C. As it happens, Princeton is located near Bell Labs, and professors on leave from there taught several of my courses. I don't recall them inveighing against formal correctness proofs, but I assume they had a subtle effect on me. Even for students not at Princeton, however, the "UNIX folks" exerted influence due to the books they wrote, not just on UNIX and C, but on other programming topics as well. I've already mentioned *Programming Pearls* by Bentley (who gave a guest lecture to us at Princeton). Kernighan, one of the authors of C, wrote a book called *The Elements of Programming*

Style with P. J. Plauger, another employee. The book is consciously modeled on William Strunk Jr. and E. B. White's *The Elements of Style*, and consists of a series of examples supporting a set of maxims, the first two of which are "write clearly—don't be too clever" and "say what you mean, simply and directly."[21]

The book is meant to support structured programming. As the preface to the second edition (1978) says about the first edition (1974), "The first edition avoided any direct mention of the term 'structured programming,' to steer well clear of the religious debates then prevalent. Now that the fervor has subsided, we feel comfortable in discussing structured coding techniques that actually work well in practice." Notwithstanding that, Kernighan and Plauger eschewed both the theory of the first *Structured Programming* book and the formal proofs of the second to focus on code itself: "The way to learn to program well is by seeing, over and over, how real programs can be improved by the application of a few principles of good practice and a little common sense."[22] It's incremental: keep improving your programs and they will wind up being good; here are seventy-seven pieces of advice on how to do that. Or consider this comment about GOTO in the original C book by Kernighan and Ritchie: "Although we are not dogmatic about the matter, it does seem that goto statements should be used sparingly, if at all."[23] This is a far cry from Dijkstra's strident denunciation.

The book *Software Tools*, also by Kernighan and Plauger, has a similarly nuanced view of GOTOs. Discussing other control structures available in the language they are using (Ratfor, short for Rational Fortran, as the name implies a more "sensible" variant of Fortran), they state, "These structures are entirely adequate and comfortable for programming without goto's. Although we hold no religious convictions about the matter, you may have noticed that there are no gotos's in any of our Ratfor programs. We have not felt constrained by this discipline—with a decent language and some care in coding, goto's are rarely needed."[24]

This message lands quite easily on the ears of a self-taught programmer. It reinforces the idea that there is not a lot of formal knowledge associated with software engineering, and your own personal experience is a valuable guide in how to proceed. Sure, you could improve a bit, but who can't? And to the extent that I picked up any sense of software engineering in college, it was this vibe. Not only did I adopt C from the UNIX crowd, but I also acquired the incrementalist view of software quality.

And on the question of formally testing your software, I acquired no knowledge at all, even from the Bell Labs visitors; it wasn't part of the curriculum at Princeton. The algorithm was what was important, and proving that you had accurately translated it into code was less so. Understand that the stakes are lower in college. When I wrote a compiler for a class, it was to learn how to write a compiler, not to use it to compile a lot of programs. If it failed on the programs I tested it with, then I fixed those issues; it was never subject to anything more stressful and at the end of the term was set aside. As Weinberg writes, "Software projects done at universities generally don't have to be maintainable, usable, or testable by another person."[25] The notion of intentionally trying to make a program break by feeding it unusual inputs or navigating the user interface in an unusual way never occurred to me. If I felt the algorithms were correct, the code looked reasonable, and I hadn't observed any crashes, then it was perfect as far as I was concerned. The programming contest that I talked about in chapter 4 presented a slight concern, since the actual data it would be run on was kept secret. I recall running it using several randomized data sets that I generated; nonetheless, when my program ran successfully (albeit not as expeditiously as some others) during the actual contest, my reaction was more "thank goodness!" than "well, of course, I tested it."

My high school friends who studied engineering at Canadian universities participated in an event called the Ritual of the Calling of an Engineer when they graduated. At this ceremony (which was designed by Rudyard Kipling, of all people, back in the 1920s), they were presented with a rough-hewn iron ring, which is worn on the little finger of their working hand and "symbolizes the pride which engineers have in their profession, while simultaneously reminding them of their humility."[26] Now it's not that I have observed any particularly greater humility or aversion to C string handling in graduates of Canadian versus US universities. Still, I left college with no sense of the impact that buggy software could have on the world. Partly this was because back in 1988, this impact was much more constrained, with computers rarely even being connected to a network. Even the Morris worm, which was released later that year, did nothing to change this attitude. My experience with the effect of bugs had been limited to my IBM PC BASIC *Pac-Man* game not working correctly or having to work late in von Neumann to get my 3-D billiards animation to display properly. Certainly as a user I had been annoyed when I lost work due to a crash in a program I was using, but I can't recall ever connecting that feeling with the notion that I would now be the one writing software that others would be entrusting their data to.

Understand that I am not blaming a UNIX slacker attitude for this; the programmers at Bell Labs obviously cared greatly about the quality of their software, since it was supporting critical telephone infrastructure. And in their books they talked about quality a lot, although the approach was incremental improvements in their software until it was of high quality. I just never got any message of "once you graduate, this stuff really matters."

So what did companies, writing larger pieces of software for customers who were paying them money, do to ensure that the software worked? Unquestionably, to the extent that they were inspired at all, it was by the UNIX approach.

At Dendrite, where I worked immediately after college, there was little testing of the software—not atypical for a small start-up that employed about ten programmers when I started. There were several customer support people working in the office, and they would run through basic operations with a new release before sending it out (by mailing a floppy disk to every sales representative running the software!), but there was no official verification process. They counted on the programmers getting things right (which to our credit, we mostly did, not through application of any formal design process, but just through being careful; the tricky bug I described in chapter 3 was unusual in that it was seen by customers).

Dendrite was small enough—all the programmers sat together in a warren of cubicles—that communication with other programmers was easy: we could yell if we wanted a question answered. When I got to Microsoft, the Windows NT team had around thirty programmers on it, laughably small compared to today, but much larger than Dendrite. In addition, Windows NT had been in development for a year and a half, so already had a sizable body of code that I needed to get familiar with.

Even at Microsoft, a well-established company at the time, there was no training on how to proceed with the task of software engineering. The general attitude was, "You're smart, so you figure it out." Trial and error was the foremost technique employed, with imitating existing code that looked similar a close second. You might wonder if this was intentional—if there was a sense that asking somebody for help was a sign of weakness. I never saw any indication that this was a policy; it was just that everyone else had learned to program by figuring things out on their own, so they simply kept doing it that way and never gave the matter much thought. Imagine if new electricians learned this way.

At Microsoft, I encountered a group of people with the title software test engineer—the *testers*—whose job it was to take the software that the developers produced and give it a stamp of approval before it was released to customers. In its early days, Microsoft had no testers either—IBM PC DOS, the software that started Microsoft on its road to prominence, must therefore have been released without a separate testing team—but eventually management realized that counting on developers to test their own code posed problems. One concern was that developers spending time testing their code was inefficient; users were more common than developers, so you could hire people to pretend to be users while the developers churned out more features. There was also this sense that you couldn't "trust" developers to test their own code; that if you asked a developer, "Tell me when your code has been adequately tested?" they would immediately respond, "It's good."

I always felt that the implied laziness/evilness/delusionality on the part of developers was unfair, and attribute their overoptimism to not having been exposed to testing in college. I found that most developers, when they arrived at a place like Microsoft, fairly quickly adopted a more conscientious approach (for example, I don't recall DOS or Microsoft BASIC being buggy). In any case, the profession of software tester was already extant at Microsoft when I arrived in 1990. The book *Microsoft Secrets*, which came out in 1995, fills in some of the background: the first test teams were set up in 1984; there were expensive recalls of two pieces of software, Multiplan (a spreadsheet, which was a precursor to Excel) in 1984, and Word in 1987; and in May 1989, there was an internal meeting on the optimistic subject of "zero-defects code."[27]

The testers would come up with *test cases*, sequences of steps designed to exercise different areas of the code; if a spreadsheet supported adding two cells together, there would be a test case to create two cells, have the spreadsheet add them, and verify that the result was correct, while also keeping an eye out for any untoward behavior, such as a crash or hang. The goal in having the test cases be formalized was to both ensure that nothing was missed and hopefully have a reliable set of repro steps for whatever went wrong.

The first edition of Cem Kaner's *Testing Computer Software*, the most recommended testing book at the time, had come out in 1988; people had been writing books about testing for at least a decade before then, and they had been testing as well as thinking about testing software for a while before that. In 1968, Dijkstra stated (in support of structured programming, remember, which advocates up-front proof rather than after-the-fact testing) that testing was "a very inefficient way of convincing oneself of the correctness of programs," and the following year he formalized this as "program testing can be used to show the presence of bugs, but never to show their absence," which he repeated in his essay in the 1972 *Structured Programming* book.[28] Mills and the IBMers, in their *Structured Programming*, state as fact (with the same motivation as Dijkstra), "It is well known that a software system cannot be made reliable by testing."[29] Well known, but not apparently to people at Microsoft a decade later, who indeed were attempting exactly that.

In chapter 2 of his book, Kaner delves into the motivation for testing; his main points are summarized in the titles of sections 2.1 and 2.2, "You Can't Test a Program Completely" and "It is *Not* the Purpose of Testing to Verify That a Program Works Correctly," respectively.[30] What, then, is testing for? Section 2.3 explains it: the point of testing is to find problems and get them fixed. Kaner's reasons for throwing cold water on your dreams of thoroughness come from G. J. Myers's 1979 book *The Art of Software Testing*: if you think your task is to find problems, you will look harder for them than if you think your task is to verify that the program has none.[31] And while you are puzzling that one out, I'll throw in this bit of existential angst from Kaner: "You will never find the last bug in a program, or if you do, you won't know it."[32]

Kaner doubles down by stating, "A test that reveals a problem is a success. A test that did not reveal a problem was a waste of time."[33] His point, also borrowed from Myers, is that a program is like a sick patient whom a doctor is diagnosing; if the doctor can't find anything wrong and the patient really is sick (and software, the analog of the patient here, is assumed to be "sick"—that is, to have bugs), then the doctor is bad.[34] Of course, the doctor finding nothing wrong after a battery of tests is different from any given test not finding anything. The baseball player Ichiro Suzuki, who is known to get a lot of hits but also swing at a lot of bad pitches, was once asked why he didn't ignore the bad pitches and swing at the ones that were going to be hits. Ichiro explained that it didn't quite work that way.

I believe what Kaner was trying to do, by focusing on finding bugs, was a shift away from the notion of "the testers said the software was good," which implies they should be blamed if it turns out it wasn't, to "the testers said they couldn't find any more bugs," which implies that we are all stepping to the same tragic pavane. Despite this, the notion of "testers signing off" was in full force at Microsoft in 1990. Which of course was another way for developers to avoid blame for bugs: it's the testers' fault for not finding them. It didn't help that Myers, back in 1979, had pushed the idea that programmers could not and thus should not try to test their own software: "As many homeowners know, removing wallpaper (a destructive process) is not easy, but it is almost unbearably depressing if you, rather than someone else, originally installed it. Hence most programmers cannot effectively test their own programs because they cannot bring themselves to form the necessary mental attitude: the attitude of wanting to expose errors."[35]

What emerged, unfortunately, was the "throw it over the wall" culture.[36] Developers would strive to reach "code complete," meaning that all the code had been written and successfully compiled, and if the stars aligned just right, might even be bug free. They would then hand the software off to the tester, with the implication that they had done their part and the tester was responsible for figuring out if it worked or not. Code complete is not an inconsequential milestone; it meant that none of your early design decisions had painted you into a corner, and the API you had designed to connect the pieces together was at least functionally adequate. But the problem was that code complete was viewed as "the developer's work is done, pending any bugs received; if the software ships with bugs, it's the testers' fault for not finding them." And woe betide the tester who pushed back, saying they could not test in the code in time; they were supposed to deal with whatever they got from the developers. At one point in the Windows NT source code, next to code for a feature that had been disabled because the testers could not find the time to test it, there existed a snarky comment from a developer along the lines of, "Well, now that the testers are designing our software for us, we'll remove this."

I don't mean to make developers look entirely bad. We do have a craftsperson's pride in their work and don't want bugs to happen; we certainly feel some sense of guilt if our programs freeze or crash, especially if they lose data. And we generally enjoy the intellectual challenge of fixing bugs. I do believe that I wrote solid code in my early days at Microsoft, but it was not because of any sense that Microsoft's stock price could be affected if I messed it up. Morris, author of the Morris worm, intended it to spread much more slowly than it did, with the goal of lurking for a while before being revealed. Presumably he was angry with himself that he had not tested the worm first to get a sense of how fast it would replicate, which would have prevented the rapid detection brought on by the havoc that it wreaked; at the time, one observer commented that he should have tested it on a simulator before releasing it.[37]

The sense that testers were there to guard against devious developers trying to sneak bugs past them often led to an antagonistic attitude between developers and testers. Just as developers viewed their users as an annoying source of failure reports rather than as the people who ultimately paid their salary—shades of the old luser attitude—they came to see testers in the same light, since their interactions with testers had a depressing similarity: every time you heard from a tester, they were reporting a failure, interrupting whatever lovely new technical problem you were working on and replacing it with a bug investigation, which always had the potential to turn into a bottomless pit. The response to a failure report was frequently a heavy sigh followed by a quick attempt to determine if the investigation could be shuttled off to another developer, and finally a "why would you use the software that way?" eye roll. Unfortunately the message that "the tester's job is to find bugs" led to using bug counts as a measure of the effectiveness of testers, which led to testers sometimes favoring quantity over quality when looking for bugs, filing many bugs for obscure cases rather than looking for problems that users were most likely to hit, which of course did nothing to improve developers' opinion of testers. But overall the bad actors here were the developers, not the testers.

Worst of all were bug reports with inexact repro steps, such that when a developer took the trouble to walk through the steps, the bug would not reproduce. Ellen Ullman's novel *The Bug* captures the general attitude here, as a programmer tries to reproduce a bug report from a tester:

He started up the user interface, followed the directions on the report: He went to the screen, constructed the graphical query, clicked open the RUN menu, slid the mouse out of the menu. Now, he thought, the system should freeze up now. But nothing. Nothing happened. "Shit," he muttered, and did it all again: screen, query, click, slide, wait for the freeze-up. Nothing. Everything worked fine. "Asshole tester," he said. And then, annoyance rising, he did it all again. And again nothing.

A kind of rage moved through him. He picked up the first pen that came to hand—a thick red-orange marker—and scrawled "CANNOT REPRODUCE" in the programmer-response area of the bug report. Then in the line below he added, "Probable user error," underlining the word "user" in thick, angry strokes. The marker spread like fresh blood onto the paper, which he found enormously satisfying ... Not a bug. They were idiots, all of them, incompetent.[38]

Ullman's book is fiction, but she is a veteran Silicon Valley programmer; the book was written in 2003, but is set in 1984, and that attitude of superiority that programmers had toward testers, possibly with a little less swearing, was still prevalent at Microsoft in the early 1990s. A programmer investigating a failure report very much hoped to reproduce it on their own machine, where they could use debugging tools to determine the fault and then work their way back to the defect in the code. If they couldn't reproduce the failure, then they would reach for the thick red-orange marker (conceptually, since even back in 1990, Microsoft used an electronic bug-tracking system) and mark it as "Not Repro," no matter how severe the effect of the failure was.

There was some software where testing was taken seriously and bugs were appreciated rather than dreaded. But this was only in situations where the developers realized that the cost of a failure would be severe—two common examples being software that ran on medical devices and spacecraft. Of course these occasionally had bugs also, such as the Therac-25 machine administering fatal doses of radiation or the Ariane 5 rocket that self-destructed (certainly not the only space mission that aborted due to a bug), but they were engineered with more care than we undertook at Microsoft. As opposed to being viewed as excellent examples of fine software craftspersonship, however, these were considered by programmers to be old, stodgy projects—imagine having to spend all that time simulating every possible situation just to be sure your software worked properly all the time!—as compared to the cool, fast work we were doing at Microsoft.

In 1990, certainly, nobody sat us newly hired developers down and talked about the expensive recalls of the 1980s. They may have harangued the *testers* about this, but I never heard about it. Nobody read us this quote, from Mills in 1976: "It is well known that you cannot test reliability into a software system."[39]

Eventually, though, the needle swung away from testers "testing in" quality and back to the notion that programmers should "design in" quality. This is the same idea that the structured programming movement was going for, but with a different approach—which I will discuss in the next chapter.

6 Objects

Imagine that the hardware store in your town was sold to a new owner. The previous owner had arranged the store with similar items grouped together: the paint over here, hand tools over there, and raw materials some place or other. It wasn't always obvious where things were, and it always seemed that if you were only buying two things for the same project, they were on opposite sides of the store, but if you knew where a certain type of merchandise was stored, you could find similar things nearby.

Now imagine that the new owner chose to reorganize the layout based on tasks. They might reason that somebody building a chair needs wood, a saw, screws, a screw gun, and paint. They group all these things together in one area and call it the "make a chair area." The people planning this would spend a while thinking through various activities they needed to support and assigning their inventory to different categories.

If you went to the store and wanted to make a chair, this would be great, since it's all in one place, nice and convenient. But what if you wanted to build a chair and you needed something that the designer of the "make a chair area" hadn't anticipated? Now you have to go hunting for the item, trying to guess what activity the new layout specialists had bucketed your missing item into.

And more fundamentally, you're trying to build a chair. Your shopping experience at the hardware store could be slightly improved or worsened by changing the layout of the store, but in the greater scheme of things, the hard part is figuring out what parts you need and actually building the chair, and that doesn't change.

To the owners of the store, though, the layout would feel logical, and they would feel they had done something clever and forward thinking. In fact, they might decide that they had revolutionized the art of hardware store design. They might even write a book and go on a speaking tour, explaining to other hardware store owners how significant their changes were. Naturally they would find a few chair builders who preferred the new layout and gather positive quotes about the experience.

The quest to "design in" software quality eventually evolved to a similar point, although the beginnings were more humble.

Wirth, creator of Pascal, published a book in 1976 titled *Algorithms + Data Structures = Programs*.[1] The title is accurate; a program consists of algorithms—the code that runs—and data structures—the data that it operates on. Both are important, and a programmer who wants to learn how a program operates must understand both of them. The "structured programming" push was aimed at clarity of algorithms, as expressed in code, but clarity of data structures received much less attention.

Data structures are built up from the fundamental data types: numbers and strings. In many early languages, there was no way to group such data together: if you wanted to store both a person's name and age, you defined two separate variables, without any visible connection between them, except whatever could be shoehorned into the names of the variables. Only by reading the code that used them could another programmer intuit that they were related. To continue my hardware store analogy, it's as if different power tools were stored randomly all around the store, because nobody had thought of a way to put them together.

The one grouping construct that existed was arrays, which were designed to hold multiple instances of the same sort of value, such as a top-ten high score list. Sometimes, for lack of any other way, programmers would use arrays for grouping different pieces of data; for instance, a BASIC program wanting to store somebody's height and weight might create an array of two integers, and put the height in the first and weight in the second. This connected them, but it was not obvious, from the declaration of a single array, what the specific meaning of each element was, and it was easy to get them backward in the code. The *Hockey* game in *BASIC Computer Games* used an array of seven strings, storing the names of the six players in the first six elements, and the name of the team in the seventh element. This made it a bit confusing to read, which wasn't helped by the fact that the array in question was named A$.[2]

Languages such as Pascal and C avoided this clunky feeling by letting you bundle related data into a larger entity, called a *record* in Pascal and a *struct* in C. For example, a C struct could hold information about a person:

```
struct person {
    int age;
    char name[64];
};
```

and if you had a variable p of type person, you could refer to the individual elements as p.age and p.name, but also refer to an entire person, such as in a function parameter. This was an underappreciated step forward in program clarity since it addressed the second part of Wirth's equation.

One of Wirth's points is that algorithms and data structures are closely related: "The choice of structure for the underlying data profoundly influences the algorithms that perform a given task." Furthermore, neither one will likely remain unchanged during the writing of a program: "In the process of program construction the data representation is gradually refined—in step with the refinement of the algorithm—to comply more and more with the constraints imposed."[3]

The person struct shown above looks reasonable—you are storing the age and name—but while working on a program with such a data structure, you may see opportunities for "refinement." You realize that the age of a person needs to be updated every year, even though their birthday will never change, so you decide to store their birthday instead of their age; this means you have to recalculate their age if you need it, but that's not too hard. Or you might notice that several times in your code, you have had to split the name into first and last names. Again, this isn't hard—hopefully the second time you did it, you moved that code into a separate API so you could call it from anywhere as needed—but it seems easier to store the first and last name separately. These are the kinds of details you might not think of when first sketching out your data structures; it's the normal evolution of a program's design.

At a mechanical level, however, this sort of change can be a pain when working in a *procedural language*—the type of language we have seen thus far, and a category that includes Fortran, BASIC, Pascal, and C. Although there is clearly a logical connection between your data structures and algorithms, they tend to be separated in the source code; the data structures defined near the top and implementation of the algorithms further down, or in a separate file. So when making improvements like this, which involve modifying both the definition of your data structures and code that uses them, you wind up moving back and forth in your source code. More important, if you wanted to figure out, say, which functions operated on a person or used the soon-to-be-replaced person.age, you would need to scan the parameter lists of all your functions to find those that took a person, and then scan the code to find any use of age.

Starting in the 1960s, the idea emerged of grouping data structures together with the code that operates on that data, in a construct known as a *class*. A class is actually a blueprint for such a collection; an instance of the data in a class, in the memory of a computer running a program, is called an *object*. As a result, this approach is labeled *object-oriented programming*.

We saw an illustration of this in chapter 3, since C# is an object-oriented language. The string class defines both data (the characters in the string) and methods (the currently preferred term for an API in object-oriented programming) that operate on the string, such as ToUpper() in our examples. You use dot notation to join the object and method name:

```
upperstring = mystring.ToUpper();
```

We are calling the method ToUpper() on the object mystring, which is an instance of the string class. The code inside the implementation of ToUpper() will be operating on the class data of that mystring object.

In a procedural language like C, you call functions directly rather than on an object; any function that wants access to data needs to have it passed as one of its parameters.[4] It doesn't look much different:

```
upperstring = ToUpper(mystring);
```

This is sometimes stated as "Rather than calling a function to uppercase the string, you ask the string object to uppercase itself," as if the objects have become self-aware and relieved their human overlords of mundane coding tasks. In reality somebody still has to write the code, but things get cleaner: all the methods that operate on the data in the class are grouped together in one place in the code, next to the definition of the data itself.

The first language to introduce classes was Simula, developed at a research lab in Norway in the 1960s (one of its authors was Ole-Johan Dahl, later the coauthor of one of the *Structured Programming* books). Simula was a general-purpose programming language, but designed with the goal of writing simulations, such as for cars on a highway or people waiting in line at a bank. The object model works quite well for simulating real-world objects: the object representing a car will have certain pieces of data associated with it, like speed and position, and will have certain operations that it can perform, such as accelerating or turning. Grouping them together in a class makes this clear to somebody reading the code.

In Simula, you would create a new object of a given class (which is also known as *instantiating* an *instance* of that class) using the keyword NEW followed by the name of the class, like this:

```
MyObject :- new MyClass;
```

where :- is known as the reference assignment operator (modern eyes should read it as an equal sign), and you could also specify that the creation of an object could take what were called *class declaration parameters*, like this:

```
MyPerson :- new Person(name, age);
```

where the Person class included initialization code, run every time that a Person object was created, that could use the name and age parameters as it saw fit. In this example, it would presumably store them in the class data so they could later be used in *class procedures* (which is what Simula called methods). This code that runs when an object is created is generally known as a *constructor*, although Simula did not use that term.

Classes can also have subclasses. This concept is now known as *inheritance*, although again, Simula did not call it that (the term inheritance, "Object B inherits from Object A," implies a more anthropomorphic view of code than the businesslike Simula phrasing "Object B is a subclass of Object A"). A subclass has all the data and procedures of the superclass, plus whatever other ones it adds. This again fits well with the simulation focus: you can have classes for animal, vegetable, and mineral, and then subclasses of animal for specific animals; you could layer it as deep as you would like.[5] A pointer that referenced an object of a specific animal class could also be used in a context (such as a procedure parameter) where the more general animal class was expected (you could also switch the pointer between referencing the specific and general classes using a keyword with the lovely name QUA, which sadly has not been picked up by any modern languages). This let you reuse common data and procedures in the animal class, but allow the specific animal classes to have their own added data and procedures. If you had a class Animal and a subclass Dog, Animal could have a procedure GetName(), which applies to all animals, and Dog could have a procedure GetBreed(), which makes sense for dogs but not most other animals.

Simula also introduced the important object-oriented notion of a virtual procedure. You want your code to be able to refer to all animals using the `Animal` class, which makes it nice and generic, but perhaps you want the implementation of `GetName()` to be provided by `Dog`, since that code would be aware of any dog-specific details. The solution is to declare `Get-Name()` as a virtual procedure in `Animal`. `Animal` can implement `Get-Name()`, but `Dog` can provide its own more specific implementation; Simula would look, at runtime, for the innermost class (that is, the deepest level subclass) with an implementation of `GetName()`, and call that one.[6]

When Simula first appeared, objects were seen as a notational convenience, not a major breakthrough in how software was written. The 1973 book *SIMULA Begin*, written by the authors of the language, doesn't use the term *object-oriented programming* at all, and presents classes and objects as just one useful feature in the language.[7] R. J. Pooley's *An Introduction to Programming in SIMULA*, which came out in 1987, doesn't get to the concept of classes or use the term object-oriented programming until chapter 9, although it does use objects in examples in earlier chapters. Pooley states, "One major advantage of this approach is that, given a sensible choice of names, we will have a much more readable program. Complicated detail is moved from the main part of the program to the procedure[s] ... of the class and replaced by meaningful procedure names."[8]

The language that popularized object-oriented programming was C++, created by a Danish computer scientist at Bell Labs named Bjarne Stroustrup starting in 1979—just a decade after C, the language on which it was based, was invented. The name is a reference to the ++ operator in C, which increments the value of a variable; Stroustrup states that the name had "nice interpretations," although he mentions that ++ can also be read as *next* and *successor*, which sounds a bit nicer then the "incremental" interpretation.[9]

Stroustrup helpfully wrote the book *The Design and Evolution of C++*, in which he explained his thinking in designing the language. Although he cleaned up a few things he disliked about C, the primary goal was certainly to bring over the class idea from Simula; his initial name for his language was "C with Classes." He beefed up Simula's class support, such as by allowing multiple constructors for a class, as long as each constructor had a unique parameter signature so the compiler could tell them apart. For that matter he came up with the term constructor, after first trying out *new-function*. Stroustrup used the terms *derived class* and *base class* instead of *subclass* and *superclass* because he thought they were clearer, with the potential confusion being that a subclass is an expansion of the superclass, not the other way around as one would expect from the way the word *subset* is used in mathematics. He also referred to the data and functions of a class as *members*, a term that has since become standard (C++ did not use the term *method* but instead called them *member functions*).[10]

More important, Stroustrup made all class members default to being *private* rather than *public*. If a variable in a class was declared as private, it was hidden from code that used that class (known as *calling* code or a *caller*); only the code inside the class itself, implementing class member functions, could access it (member functions themselves could be similarly hidden, which restricted them to being called only by other member functions). In the first version of Simula all members were public; in the early 1970s the notion of private members was added, but the default was still public—accessible directly by any caller unless you explicitly marked members as private in your class declaration.[11] C++ made the opposite choice, with members being private unless otherwise indicated. This encouraged *abstraction* of the implementation from callers; implementation details can change without affecting code that calls the class, as long as the only changes are to private data and functions, and the public surface remains the same.

David Parnas is a computer science professor who wrote the first papers on "information hiding," as he called abstraction: the idea that keeping different modules from knowing the internal details of each other's data structures made them more robust. As he observed, "It was information distribution that made systems 'dirty' by establishing almost invisible connections between supposedly independent modules."[12] In his original 1971 paper on information hiding, Parnas wrote about the connections between modules:

Many assume that the "connections" are control transfer points, passed parameters, and shared data. ... Such a definition of "connection" is a highly dangerous oversimplification which results in misleading structure definitions. The connections between modules are the assumptions which the modules make about each other. In most systems we find that these connections are much more extensive than the calling sequence and control block formats usually shown in system structure descriptions. ...

We now consider making a change in the completed system. We ask, "What changes can be made to one module without involving changes to other modules?"

We may make only those changes which do not violate the assumptions made by other modules about the module being changed. In other words, a single module may be changed only as long as the "connections" still "fit." ...

The most difficult decisions to change are usually the earliest. The last piece of code inserted may be changed easily, but a piece of code inserted several months earlier may have "wormed" itself into the program and become difficult to extract.[13]

In a 1972 paper, discussing the division into modules of a larger program, Parnas summarizes the approach: "Its interface or definition was chosen to reveal as little as possible about its inner working."[14] This gave the owner of a module the maximum flexibility to rework it without having to modify all the code that called its API.

Based on my experience in college working on programs involving a maximum of two programmers, I was unaware of these finer points and would not have understood why they mattered. Nonetheless, C++ began to catch on as the next logical successor to C. I believe the first I ever heard of the language was from a classmate at Princeton, probably in early 1988, describing a program they wanted to write and saying something along the lines of, "A few years ago I would have written it in C, now of course I would write it in C++." He was probably just showing off, but still it demonstrates how the language was gradually intruding into the consciousness of programmers as the new thing. Nevertheless, the original name, C with Classes, is indicative of what Stroustrup was aiming for: take the language C and add the useful feature of classes, as opposed to changing the world through object-oriented programming. In the preface of the original C++ *Programming Language*, he states (using the term *type*, where a class is a user-defined type), "In addition to the facilities provided by C, C++ provides flexible and efficient facilities for defining new types. A programmer can partition an application into manageable pieces by defining new types that closely match the concept of the application. ... When used well, these techniques result in shorter, easier to understand, and easier to maintain programs."[15] Stroustrup doesn't get to an in-depth discussion of classes until chapter 5, toward the middle of the book.

There was another object-oriented language floating around at the time called Smalltalk, which had been developed at Xerox's Palo Alto Research Center (PARC) during the 1970s. The first general release was called Smalltalk-80, after the year in which it was appeared. The book *Smalltalk-80: The Language*, coauthored by Adele Goldberg, one of the designers of the language, and David Robson, explains, "The Smalltalk-80 system is based on ideas gleaned from the Simula language and from the visions of Alan Kay."[16] Kay is a good source for visions. If you have heard of Xerox PARC, it is likely as the place that invented graphical windowing environments, later appropriated by Apple and Microsoft; Kay was one of the leaders on that project, in addition to coining the term *object-oriented programming*.

Stroustrup notes that he had heard of Smalltalk when he was designing C with Classes, but doesn't list it as a primary influence (Smalltalk-80 was the fifth version of the language; the language predates Stroustrup's work adding classes to C).[17] In Smalltalk you don't call methods, you send messages to an object, which that object may choose to process by calling a method (so a method is an internal implementation detail in a class, not the public surface; another effect of this is that all class members are necessarily private). Smalltalk is a "pure" object-oriented language; even language constructs are based on objects. Instead of an IF/ELSE statement, as in most languages, you have an expression that evaluates to a Boolean object (an object that stores a value that is either true or false), which is then sent a message with two other objects containing blocks of code (because code itself is also an object), one of which should be run if the Boolean is true, and the other which should be run if the Boolean is false, like this:[18]

```
number > 0
    ifTrue: [positive ← 1]
    ifFalse: [positive ← 0]
```

This is described in Smalltalk terminology as sending a message with the *selector* ifTrue:ifFalse:, meaning that the message has two arguments, named ifTrue: and ifFalse: (true camel casing, you will observe). The message is sent to the Boolean object that results from evaluating the expression number > 0. This is the rough equivalent of calling a method on the Boolean that takes two parameters, except note that the parameters are identified by a name, not a position in an argument list, which also means that the syntax easily supports making them optional (if there is no ifFalse: argument, for example, it is like having no ELSE block on an IF statement). Named versus positional parameters is a minor detail at this point in our story, but keep it in mind for later.

It's all quite mind expanding, if a bit hard to read for somebody used to almost any other language. Smalltalk is proud and unapologetic in its stance; *Smalltalk-80: The Language* jumps right into objects, messages, classes, instances, and methods in the first chapter. To its credit, it avoids ascribing hyperbolic benefits to this system, merely stating, similar to what Stroustrup said about C++, that "an important part of designing Smalltalk-80 programs is determining which kinds of objects should be described and which message names provide a useful vocabulary of interaction among these objects."[19]

Given that C++ was emerging at around the same time as Smalltalk-80, with its similarity to C making it appealing to people who had been seduced by that language, it is understandable that Smalltalk, whose syntax was unusual for anybody familiar with the Algol-Pascal-C language family (which is to say, almost everybody), never gained as much mainstream traction as C++.

Programmers began playing around with C++, especially C programmers who wanted to try something slightly edgier; they happily divided their class variables into public and private, enjoying the excitement of playing with their new object-oriented toys.

In 1986, the first Object-Oriented Programming, Systems, Languages, and Applications (OOPSLA) conference was held in Portland, Oregon, under the aegis of the Association for Computing Machinery (ACM). Goldberg was one of the organizers. OOPSLA is one of a series of specialized conferences that the ACM puts on (in 2010, the conference was merged into the Systems, Programming, Languages, and Applications: Software for Humanity [SPLASH] conference). Stroustrup describes the first OOPSLA conference as the "start of the OO [object-oriented] hype."[20]

OOPSLA arrived at the beginning of the third act of an important arc in programming language design. Although IBM was behind the creation of Fortran and PL/I, many of the first computer languages were developed at universities: BASIC was invented by two Dartmouth professors, Kemeny and Kurtz, Pascal by Wirth at the ETH in Zurich, and Algol by a committee of computer scientists. These were simpler days, where languages and the problems they solved were much less complex than they are today, and a better match to the capacity of a college professor. After that we moved into an era where languages emerged from research labs, either private ones like the Norwegian lab that came up with Simula or more commonly research labs within larger hardware companies: C and C++ came from Bell Labs, and Smalltalk from Xerox PARC. These languages were invented, in various degrees, to support the company's business, but they were a by-product and not the end goal.

In the 1980s, there began to emerge languages designed by companies that were an end unto themselves: the company's business was the language, and the success of the company depended on programmers adopting a new language—always a difficult sell to programmers. Two of the first of these were Objective-C, invented by Brad Cox and Tom Love at a company called StepStone, and Eiffel, invented by Bertrand Meyer at a company called Eiffel Software. Both of these were object-oriented languages.

I am not impugning the motives behind Objective-C and Eiffel; the object-oriented approach was genuinely seen as a road to software that was both higher quality and easier to write. Nonetheless, one can appreciate that if the foundation of your business involves convincing programmers, enamored of C, to switch to an object-oriented language, you need to present object-oriented languages as a bit more than the mere notational convenience that Simula and C++ were aiming for.

Cox's book *Object-Oriented Programming: An Evolutionary Approach* was published in 1986. Despite the subtitle, his approach was evolutionary only in the sense that his language, Objective-C, was based on C rather than being completely new. In the preface, Cox gets right to it: "It is time for a revolution in how we build software, comparable to the ones that hardware engineers now routinely expect about every five years. The revolution is object-oriented programming." He does dial down the rhetoric in the body of the book; Objective-C uses message passing instead of method calls, the same as Smalltalk, which does give it more flexibility than C++ in how objects can choose to support a message. Cox at one point calls this "the only substantive difference between conventional programming and object-oriented programming."[21]

Such restraint was missing in the writings of Meyer (who presented the paper "Genericity versus Inheritance" at the first OOPSLA).[22] He starts his 1988 book *Object-Oriented Software Construction* with this paragraph, which rivals anything Dijkstra ever wrote for sheer awesomeness:

Born in the ice-blue waters of the festooned Norwegian coast, amplified (by an aberration of world currents, for which marine geographers have yet to find a suitable explanation) along the much grayer range of the Californian Pacific; viewed by some as a typhoon, by some as a tsunami, and by some as a storm in a teacup—a tidal wave is reaching the shores of the computing world.[23]

Following on to the oblique references to Simula and Smalltalk, the next paragraph acknowledges that the reader may have heard this sort of thing before: "'Object-oriented' is the latest *in* term, complementing or perhaps even replacing 'structured' as the high-tech version of 'good.' ... Let's make it clear right away, lest the reader think the author takes a half-hearted approach to this topic: I do not think object-oriented design is a mere fad." Meyer then throws down the gauntlet: "I believe it is not only different from but even, to a certain extent, incompatible with the software design methods that most people use today—including some of the principles taught in most programming textbooks. I further believe that object-oriented design has the potential for significantly improving the quality of software, and that it is here to stay."[24]

He states that he will show

how, by reversing the traditional focus of software design, one may get more flexible architectures, furthering the goals of reusability and extendibility. ... When laying out the architecture of a system, the software designer is confronted with a fundamental choice: should the structure be based on the actions or on the data? In the answer to this question lies the difference between traditional design methods and the object-oriented approach.[25]

He presents the top-down functional approach as the old way, in which you start with the overall goal of a program and then break it down into smaller pieces of functionality; this is what structured programming was about. This, he argues, tends to lock the software into a certain functionality, making it hard to modify when user requirements (inevitably) change, and preventing reusability, since the pieces that the software is broken down into will be specific to the overall function (which is ironic, since *SIMULA Begin* presents both top-down and bottom-up routes to the same goal, and in *An Introduction to Programming in SIMULA*, the authors talk about "how object-oriented programming makes top-down design easier," because you can rough out your object interfaces without worrying about implementation details).[26]

Meyer then explains how object-oriented design avoids these problems, because your data tends to change less than your functionality and can more likely be reused in other components. He states the mantra, "Ask not first what the system does; Ask WHAT it does it to!" and then posits, "For many programmers, this change in viewpoint is as much of a shock as may have been for some people, in another time, the idea of the earth orbiting around the sun rather than the reverse."[27]

As with structured programming, object-oriented programming could be viewed as either a process to arrive at an object-oriented program or the resulting program itself. The initial discussion, in the days of Simula and Smalltalk, was about the resulting program: it had objects, so it was object oriented. Meyer is instead talking about object-oriented design—a process that begins before you start writing the code, thereby representing a more fundamental shift in approach.

Indeed in his view, object-oriented design does not need language support; his book includes a section on how to write object-oriented code in existing languages, including a particularly forced attempt to do it in Fortran, although he washes his hands of Pascal.[28] Nonetheless, Meyer is clearly claiming that object-oriented design turns out better when programming in an object-oriented language, and a lot of what he talks about, such as allowing classes to be extensible (which is done through inheritance), does require language support. His basic claim is that bottom-up design, in which you start by defining your classes and then stitch them together into a program, produces designs superior to those produced by top-down design.

The way in which the design of a program actually evolves is not as simple as Meyer implies. You create a class, defining your best guess as to what the methods should be. Later you might realize that the code that calls those methods needs the data in a different format or needs access to details about an object that you haven't exposed; alternately, you realize that a method you have defined isn't being called by anybody, so isn't needed (for now, anyway—let's see what your code reviewer thinks). This is not bad; it's the way program design evolves. But it's the sort of driven-from-above change that object-oriented design was supposed to avoid, per Meyer.

A 2010 paper by Linden Ball, Balder Onarheim, and Bo Christensen compared breadth-first design (proceeding from the top down in an ordered way) to depth-first design (digging into specific areas, such as a single class). It summed up a point made in an earlier study, co-authored by Ball, this way:

experts will often tend to *mix* breadth-first and depth-first design. ... [T]he preferred strategy of expert designers is a top-down, breadth-first one, but they will switch to depth-first design to deal strategically with situations where their knowledge is stretched. Thus, depth-first design is a response to factors such as problem complexity and design uncertainty, with in-depth exploration of solution ideas allowing designers to assess the viability of uncertain concepts and gain confidence in their potential applicability.

The authors then verified this by analyzing video recordings of three different teams of actual programmers at work:

All design teams rapidly produced an initial "first-pass" solution ... indicative of breadth-first solution development. ... High-complexity requirements were subsequently dealt with much earlier in the transcripts in comparison to intermediate- and low-complexity requirements. ... This finding was generalized across all three design teams and suggested the use of a depth-first strategy to handle high-complexity requirements and a breadth-first strategy to deal with low- and intermediate-complexity requirements. ... Overall, these findings point to a sophisticated interplay between structured breadth-first and depth-first development in software design.[29]

In other words, programmers can go back and forth, identifying broad areas and then digging into the details when they recognize that the solution for a given area is unclear. But even for simple classes, the design does not radiate outward from the class definition in a brilliant beam of clarity; it is a dance between the providers (the class) and consumers (the callers of the class) until it settles on something that seems reasonable. It's similar to the dance between algorithm and data structures that Wirth was describing in the quote at the beginning of this chapter. To paraphrase a military saying, "No plan survives contact with the callers of your class methods." The unfortunate fact is, it's hard to know if a design is "good" until the whole program is written and working. There is no reliable process, top down or bottom up, to arrive at this situation in a deterministic way, and even if nirvana is reached, it is only a temporary respite until the requirements of the program change.

Remember the discussion of the importance of API design in chapter 3? A class is providing an API, and its design needs careful attention, like any other API. As Joshua Bloch recommended in a talk on the subject of API design, "Write to your API early and often."[30] In other words, you need to consider an API you are providing from the caller's view before deciding that it is well designed. As the caller of methods in object-oriented code, you still wind up being ruled by whatever the author of the object chose to expose. If your notion of how an API should be structured is in sync with the person who wrote the API, then it will appear obvious and intuitive, object oriented or not. And if you're not in sync, it will be puzzling and hard to work with.

At one point at Microsoft, I worked on a product that supplied an API to programmers outside Microsoft. Somebody on the team commented that the API looked reasonable, but we couldn't be sure until a lot of people had used it. My instinctive reaction was, "It's a good API. What do we care what other people think?" ... but I now see the error of my ways. Henry Baird, one of my on-loan-from-Bell-Labs professors at Princeton, points out that API design requires social skills, because you have to be aware of what assumptions your callers may make. Such empathy can be rare in programmers, who, to quote Baird again, "are strongly attracted to the idea of going into a room alone with the machine and getting something beautiful"—beautiful, of course, in their own eyes.[31]

As Stroustrup cautions, "Remember that much programming can be simply and clearly done using only primitive types, data structures, plain functions, and a few classes from a standard library. The whole apparatus involved in defining new types should not be used except when there is a real need."[32] Rushing to define your own classes doesn't have much point if you don't even need any new classes, but of course if you define your classes first, you might never realize that they are unnecessary.

Meyer at one point states, "We have seen that continuity provides the most convincing argument: over time, data structures, at least if viewed at a sufficient level of abstraction, are the really stable aspects of a system."[33] The problem is that the sufficient level of abstraction needed for this to be true is so high level that it only applies to the basic broadbrush design of a system: "We'll need a place to store the data," or "The images should be accessible in a standard way." Once you start diving in, things become much less straightforward, and you won't know your class design is right until you have written the code for all the actions you need to perform.

If you read through papers presented at the OOPSLA conferences, they are all full of interesting ideas about object-oriented programming—many of them claiming to be in the service of writing software that demonstrates desirable attributes, such as modularity, composability, reusability, and so on. They rarely, however, have research to back up these claims; they simply state that such-and-such arrangement of objects is pleasing to the eye. There is little side-by-side investigation of the "old way" and "new way," and no metrics to evaluate if arranging your objects in a certain way produces fewer bugs or more maintainable code. To the extent that there *were* papers of this sort, they came from academia or corporate research labs, especially Hewlett-Packard and IBM.

The fact that it was old-school hardware companies doing the empirical studies is not surprising. First of all, they had the capacity to do so; there were no large software-only companies in the 1980s (Microsoft Research was founded in 1991). Second, in the world of hardware, which is driven by research-based science, dramatic improvements do come from research labs. Back in 1965, Gordon Moore, one of the cofounders of Intel, predicted that the capacity of integrated circuits—the basic building blocks of computers—would double every year.[34] The timeline for doubling has since moved back to two years, more or less, but the law has proved remarkably durable for half a century. These advances were based on scientific research into the materials and processes used to fabricate integrated circuits. It would be reasonable for a hardware company, observing the term software engineering, to think that similar advances could be made on that side; software companies would be too pessimistic to try. In the late 1980s, John Young, the CEO of Hewlett-Packard, announced that "software quality and productivity had to rise by a factor of ten in five years."[35]

Trying to impose a Moore's law equivalent for software turned out to be impossible, but it did drive some good research (Brooks, who managed hardware teams at IBM before moving over to software, points out that proximity to hardware gives software a bad rap, like an extremely attractive friend who makes you feel inadequate: "The anomaly is not that software progress is so slow but that computer hardware progress is so fast. No other technology since civilization began has seen six orders of magnitude price-performance gain in 30 years").[36] Hewlett-Packard did make improvements in software productivity, as documented in Robert Grady's book *Successful Software Process Improvement*. The goal was a factor-of-ten reduction in postrelease defect density as well as the total number of open serious and critical defects. They did reach one-sixth of the previous defect density, short of the goal but still an impressive result. Unfortunately the amount of software in their products grew so much that the other metric, open and serious critical defects, remained level.[37]

The academic papers acknowledge that they report on only one study, and that more study is needed, although oftentimes this subtlety is lost. Consider the Law of Demeter, a well-known object-oriented rule that was first presented by Karl Lieberherr, Ian Holland, and Arthur Riel, professors from Northeastern University, at OOPSLA in 1988.[38] It is aimed at reducing coupling between classes—that is, how much any class knows about another class—in support of the oft-desired goal of making it easier to rework a class without breaking other code that uses that class. Specifically, the approach is to reduce the number of classes that any given class knows anything at all about (Demeter is the Greek goddess of the harvest; the law was not directly named after her, but took the name from a tool, also called Demeter, that was used for formally specifying class definitions—a rich bounty of class definitions, presumably—and was also developed at Northeastern). The Law of Demeter states that if a certain class C is using an object A, it should be oblivious about any objects returned by methods on A—*oblivious*, in this case, meaning that if code in C calls a method on A that returns an object of class B, then C should not call any methods (or access any public data members, if such exist) on B. The most C can do is hand B back to another method on A if it is needed. The law can be paraphrased as "don't talk to strangers."

The effect of this is that if B changes in any way, even in the names or parameters of its public methods, C won't need to change, because it is treating B as a black box. C only leverages knowledge of A, not B. This is stronger than saying "C only accesses public members of B," the basic coupling reducer in object-oriented programming. It's saying that C doesn't access anything in B at all.

This certainly does reduce coupling at the class level: class C is only sensitive to changes in class A, not class B. The problem is, what if C wants to accomplish something that is best handled by a method on B—you call a method on A, it returns an object of class B, now you want to use that B object to do something else? The answer, per the Law of Demeter, is that A should add a new method that provides that functionality, which it (presumably) implements by turning around and calling B internally; C is allowed to call this new method on A, since it is already coupled to A.

This is following the letter of the law, but not the spirit. C is still not coupled to B, but it is now more tightly coupled to A because it is now calling another method on A. And A is now slightly more coupled to B because it is now providing a method that it likely implements by calling B. Yes, technically, A could continue to support this new method without calling B, since this is an internal implementation detail, but that would likely be more work.

The authors of the original OOPSLA paper had been using the law in a large programming project with their students, so they had experience to back up their proposal. They do avoid extravagant claims, and acknowledge, as you would expect academics to do, that there are potential issues. The method count on a class may increase: "In this case the abstraction may be less comprehensible, and implementation and maintenance more difficult." The authors continue in the same vein: "We have seen that there is a price to pay. The greater the level of data hiding, the greater the penalties are in terms of the number of methods, speed of execution, number of arguments to methods and sometimes readability of their code."[39] They end with a recommendation for further investigation.

I don't doubt that the Law of Demeter is helpful in certain situations. What is not known, because it has not been studied formally, is what those situations are: what it is about the programming task at hand, the size of the team, the likelihood of future changes, and so on, that makes applying the Law of Demeter a net benefit. Yet, in the time since the original paper was published, the Law of Demeter has been picked up and is now presented as universally applicable object-oriented canon; I was taught it as such in an object-oriented programming class I took. I suppose "The Possibly Useful Idea of Demeter" doesn't sound as appetizing.

Some OOPSLA papers did involve solid research. In a paper presented at OOPSLA 1989, Mary Beth Rosson and Eric Gold from IBM Research point out,

A widely held belief about object-oriented design (OOD) is that it allows designers to model directly the entities and structures of the problem domain. ... This is an inherently *psychological* claim, with psychological consequences: a design approach that more directly captures the real world should ease the cognitive process of mapping from the problem to a solution, and it should produce design solutions that are more comprehensible in terms of the problem domain. Surprisingly, though, virtually no psychological analyses of such claims exist.[40]

The authors then proceed to compare actual object-oriented programmers to procedural programmers as they analyze a problem and talk through a solution. It's good stuff, but unfortunately rare for an OOPSLA paper.

Even better is a 1991 paper titled "An Empirical Study of the Object-Oriented Paradigm and Software Reuse." It starts with the obligatory group mea culpa:

While little or no empirical validation exists for many of software engineering's basic assumptions, the need for scientific experimentation remains clear. ... The use of precise, repeatable experiments to validate any claim is the hallmark of a mature scientific or engineering discipline. Far too often, claims made by software engineers remain unsubstantiated because they are inherently difficult to validate or because *their intuitive appeal seems to dismiss the need for scientific confirmation.*[41]

The study, though, is great; done by John Lewis, Sallie Henry, Dennis Kufara, and Robert Schulman, all professors at Virginia Tech, it involved a carefully planned experiment comparing reuse between procedural and object-oriented languages, using students as the test subjects. The professors had a control group. They balanced out the skill set among the students. One of the professors was a statistician! They conclude that object-oriented languages promote software reuse more than procedural ones; I could push back on the grounds that the languages used were C++ and Pascal, which isn't a fair fight, but that would be a quibble.[42] Instead, I will salute this paper as a shining beacon of engineering research.

Most of the rest, however, is anecdotal reporting on what the authors accomplished. As Marvin Zelkowitz wrote in 2013,

The typical conference proceedings today in software engineering contains numerous papers of the form

How <my acronym>, using <this new theory of mine>,

is useful for the testing of <application domain>

and is able to find <class of errors> better than existing tools.[43]

Since every software project is unique, and problems tend to show up as the group of programmers changes or the software evolves over the years, the successful completion of a project is not an indicator that the methodology and language chosen were the best ones possible. I successfully wrote games in IBM PC BASIC; that doesn't mean that it's a great language. And the wide universe of software makes it easy to construct specific examples where a given arrangement of classes in an object-oriented program works well; that doesn't mean the advice can automatically be generalized. When writing a game on the IBM PC, a lot of the code involves changing the image displayed on the screen. In this environment, making the screen globally accessible from anywhere in the program rather than requiring a screen handle or screen object is a convenience; it avoids having to pass around a parameter that will always be the same. I could have written a paper on "The Effective Use of Global Variables to Optimize Interactivity"; that doesn't mean that global variables are always good. There are many other areas, even in the code for the same game, where they make the code hard to read, not to mention hard to modify without breaking anything.

Certainly OOPSLA had a lot of "hype," as Stroustrup called it. In the abstract for a panel on "OOP in the Real World" at OOPSLA 1990, a description of a troubled project included this:

These were by and large, failures of management and many of them were quite independent of the use of an OOPL [Object-Oriented Programming Language]. Nonetheless, the fact that we were using an OOPL was important because it contributed to an attitude that would not otherwise have existed. It was very true and is still somewhat true that OOP protagonists are true believers. The very real benefits of using OOP are presented in a very one-sided fashion which too often leads to the view that OOP is a panacea. This better than life outlook induced a euphoria in management which caused suspension of the normal procedures and judgment criteria.[44]

It's not that these object-oriented ideas are bad; many of them are good. They may produce more readable, maintainable programs. Maybe all of them do! But more evidence would be helpful in knowing what actually is better, as opposed to just sounding better, and in figuring out when a particular approach will work best. I think it comes back to the fact that many programmers are self-taught; they are used to their own experiences being all the evidence they need that an idea is worthwhile. As with any situation where code is calling an API defined by somebody else, it's similar to the hardware store redesign: if your task lines up with what the designer had in mind, it can be nice, but otherwise it can make your work difficult.

Object-oriented programming has been discussed enough that the term has filtered into the public's consciousness as something that programmers do, although to the extent that it is covered in the mainstream press, it's not about improved design but rather reusability. In particular, it concerns the idea that now that you have these objects, you can easily glue them together to make programs. This perception is not surprising. First of all, the word *object* makes it sound like you can do that. Second, it's a convenient way to explain to a nonprogrammer what is new about objects compared with the old way; they are standardized components that can then be used to assemble larger pieces. Most important, object-oriented proponents repeatedly made this claim. Cox wrote that programmers would "produce reusable software components by assembling components of other programmers. These components are called Software-ICs to emphasize their similarity with the integrated silicon chip, a similar innovation that has revolutionized the computer hardware industry."[45]

This ignores a couple of problems. You could do that with procedural programming (all programs, object oriented or not, are built up in layers that have to connect with each other and interact cleanly, and "reuse" just means that somebody else wrote some of the layers, which is independent of object-oriented programming).[46] The reality is that you can slap objects together with other objects *if* they were designed to do that or happen to mesh together well, and you can't if they weren't, and that is also the same as procedural programming.

It turns out that small problems can trip up the combining of objects. Hewlett-Packard researcher Lucy Berlin's paper at OOPSLA 1990 titled "When Objects Collide" observed that "pairs of *independently sensible* pragmatic decisions can cause fundamental incompatibilities among components."[47] In other words, the designers of the code that is calling a class, and the designers of that class itself, can each make completely rational, sensible decisions about how their code works, in out-of-the-spotlight areas such as how they handle errors and how objects are initialized, which turn out to be fundamentally opposed and make piecing together objects impossible.

There has been one instance when the "stick the blocks of code next to each other and it will work" approach was successful, and it predates object-oriented programming. Back in the 1970s, the UNIX operating system introduced the notion of a *pipeline*: taking the output of one program and sending it to another program as the input. This is most accessible to users when they are using the command-line interface, which is what DOS looked like: the operating system displays a prompt and blinking cursor, the user types a command and hits enter, the output of the command scrolls past, and then the computer prompts for the next command. Although they are somewhat hidden, command prompts still exist in both Windows and macOS (not to mention Linux) because they allow certain complicated commands to be typed easily. It's not just programmers who appreciate this; in 1999, the science fiction author Neal Stephenson wrote a long paean to the power of the command line titled *In the Beginning ... Was the Command Line*, which was later reprinted as a book.[48] Kernighan and Plauger's book *Software Tools* is primarily about the usefulness of command-line tools to programmers.

Using simple syntax on the UNIX command line, you could print the contents of a file, extract data from it, rearrange that data as needed, and build useful larger "programs" out of smaller building blocks without needing to modify the underlying code—the putative benefit of object-oriented programming. You put together a series of commands using the vertical bar as the pipeline symbol, like this (which I've split to fit on the page, but in reality would be typed on one line):

```
cat filename.txt | grep total | cut -d , -f 5 |
     tr a-z A-Z | sort | uniq
```

This says "take the contents of `filename.txt`, grab only the rows that contain the word 'total,' interpret it as a comma-separated list and take the fifth column, uppercase the values, sort them, and remove duplicate rows." This is quite powerful for a lot of simple manipulation of data. It's part of the "UNIX Philosophy," which is described in Brian Kernighan and Rob Pike's 1984 book *The UNIX Programming Environment*:

Even though the UNIX system introduces a number of innovative programs and techniques, no single program or idea makes it work well. Instead, what makes it effective is an approach to programming, a philosophy of using the computer. Although that philosophy can't be written down in a single sentence, at its heart is the idea that the power of the system comes more from the relationship among programs than from the programs themselves. Many UNIX programs do quite trivial tasks in isolation, but, combined with other programs, become general and useful tools.[49]

This is a different UNIX-related notion from the "an incremental approach to program improvement is better than a heavy-handed process" one that I discussed in the previous chapter, although it could be viewed as a different side of the same coin: many small things are better than one big thing.

The reason you could stitch the command-line tools together so well was that when transferring data between them, they broke the data down into the simplest, most portable format—everything became text strings—so the result was not particularly fast, and it required the user to have knowledge of the data format. But the user could easily modify the data, precisely *because* it was just text strings; in fact, a lot of the command-line tools existed solely to manipulate the data so it could be successfully passed into the next tool, such as `cut` and `tr` in the example above. If one tool output the data separated by commas but you needed it separated by tabs, or you needed to sort the output or remove duplicate lines, simple tools were available for that.

When linking objects together you run into problems, because the output of one method can't always be easily fed into the next method; the UNIX command-line environment let you first notice the equivalent problems by visually inspecting the output of the first tool, and then fix it up as needed by inserting commands into the pipeline before feeding it into the second tool. Of course all this was slow, with all the conversion to strings and back, so stitching together commands this way wasn't considered "real" programming, but it did work.

But so far, this is the only case where the "objects as building blocks" idea works. Early object-oriented writers recognized this. Cox called UNIX command-line pipelines "one of the most potent reusability technologies known today."[50] Meyer mentions them when talking about composability, although for him that is just one of the criteria for good design: "This criterion reflects an old dream: transforming the software design process into a construction box activity, whereby programs would be built by combinations of existing standard elements."[51]

Still, in most cases objects can't be arbitrarily glued together. What can they be used for? In fact, there is a situation in which objects are unquestionably a step forward, but it's not the elegant designs that Meyer had in mind; it's more mundane, yet also more useful.

7   Design Thinking

Inheritance was one of the signature features of object-oriented programming; many new C++ programmers latched onto the notion of sharing common code in a base class. It's an elegant way to avoid having to write (and test) the same code twice. And analyzing code for commonalities, to find the logical split into base class and derived class, is the type of clever thing that programmers like to do, even if it was possible that nobody would ever instantiate an instance of the base class on its own.

Naturally this can be taken too far. I worked on a C++ project in the mid-1990s where one of the developers had decreed that any inheritance could add only one new member. If you started with a rectangle base class, and wanted to add both text and a color as members, you had to first create a derived class that added the text member, and then derive a class from that which added the color member—and most classes have a lot more than three members. God forbid that somebody decided they wanted to use a class that was a rectangle with text and the color got dragged along too. Left unanswered was, What if they wanted a rectangle with a color but no text? Anyway, limiting inheritance to one new member made no sense because … just trust me, it made no sense. I mean, does it sound like it makes sense? But it had been mandated at some point in the past. Given the usual problem that it is impossible to convince a programmer they are wrong, and the fact that the code was already structured that way, I accepted my fate.

There are other ways that programmers can overuse a cool new feature like inheritance. Consider an e-mail program that wanted to support encrypted e-mails—encoding an e-mail so that only certain people can read it. There are various encryption algorithms available, each with trade-offs in how fast and secure they are. Let's say this e-mail program wanted to use the encryption algorithm known as Advanced Encryption Standard (AES).

Further, assume that a class had already been written that provided AES encryption, defined something like this:

```
class AESEncrypter {
    public Encrypt() { }
}
```

This is what class definitions look like in most object-oriented languages: the word `class` followed by the name of the class and then the methods. Here the class is named `AESEncrypter`, and it has a single public method named `Encrypt()` (this example leaves out a few details, especially that the `Encrypt()` method presumably takes several parameters and may return a value, and also that the implementation, between the `{ }`, is empty).

The e-mail program is planning to use `AESEncrypter` to perform encryption, but there is flexibility in exactly how it goes about this. One approach, for programmers under the initial spell of object-oriented programming, was to leverage the fact that a class that inherits from a base class can access all the members (data and methods) of the base class. You could have the e-mail class inherit from the encryption class, as indicated by the colon syntax in the declaration of your new `Emailer` class:

```
class Emailer: AESEncrypter {
    // code in Emailer can call Encrypt()
}
```

and now all the methods on `AESEncrypter` are available to `Emailer`.

This isn't particularly logical, since inheritance is about extending the functionality of the base class and offering the combined functionality to callers of the derived class; your `Emailer` class exposes the public methods of any class it inherits from, which means that other code might start using `Emailer` for encryption by calling the `Encrypt()` method, thereby locking `Emailer` into providing encryption for the foreseeable future. This goes against the idea of loose coupling of classes.[1] Yet programmers would design their code this way, because it solved their immediate problem of needing access to encryption functionality and they weren't thinking about the longer-term effects of how their callers would invoke their class.

The test for whether inheritance made sense was eventually formulated as "is-a" versus "has-a." In order to inherit from a class, you should legitimately be that class—the [derived class] "is-a" [base class]—with a little something extra added. It makes sense for a `Dog` class to inherit from an `Animal` class, because a dog is an animal. Is it accurate to say that `Emailer` is an `AESEncrypter`? No, `Emailer` is for sending e-mail; it's only providing encryption accidentally, because of the way you have used inheritance. This pushes it to the "has-a" category—`Emailer` has an `AESEncrypter` that it uses. Instead of inheriting from `AESEncrypter`, it should contain an instance of it:

```
class Emailer {
    private AESEncrypter aes;    // can call aes.Encrypt()
}
```

and it can now call methods on that `AESEncrypter` as it sees fit. That `AESEncrypter` object `aes` is private—meaning its methods are not exposed to callers of `Emailer` but instead are only for use internally by `Emailer` methods.

This is cleaner than inheriting from `AESEncrypter`, but it still has the potential to cause friction in the yet-to-be-revealed future, the bête noire of all software design. As a code reviewer might ask, What if you decided that you wanted to change your encryption algorithm?[2] Or what if you wanted to support more than one encryption algorithm?

One solution would be for `Emailer` to contain multiple encrypters and use only the one it needed when sending an e-mail, but that is cheesy for several reasons. First, you would have all those encrypters lying around, sitting there in your class definition like so many bumps on a log, each taking up a small bit of memory, although only one of them was being used in a given instance of your class. Second, you would wind up writing code like this in `Emailer` to figure out which encrypter to call (based on a variable, such as `encryptionType` used below, that defines which encryption type you actually want to use):

```
if (encryptionType == AES) {
    aes.Encrypt();
} else if (encryptionType == RC4) {
    rc4.Encrypt();
} else if (encryptionType == TripleDES) {
    tripledes.Encrypt();
}
```

and similar code would need to be repeated every time you wanted to call the encrypter. Plus, all those sections of code would need to be extended via yet another ELSE IF if you added another type of encryption.

Fortunately, around this time people were figuring out an extremely useful application of object-oriented programming, involving *interfaces*.

A class can declare a method as abstract; this means that the class does not implement the method but simply sets the *contract*—the method name and parameter list—that derived classes must follow when they implement that method. Classes with abstract methods can't be instantiated as objects; only a derived class that implements all the abstract methods of its base classes can be instantiated (such a class is known as a *concrete* class).

An interface is a class that takes this to the extreme: all of its methods are abstract. Because of this you can't instantiate an instance of an interface; they exist only as contracts to be followed by derived classes that inherit from them:

```
interface MyInterface {
    int SomeMethod(string a, int b);
}
class MyConcreteClass: MyInterface {
    public int SomeMethod(string a, int b) { }
}
```

Notice that the parameter signature of SomeMethod() in MyConcreteClass exactly matches the one in MyInterface (the concrete class has to explicitly define SomeMethod() as public, which is automatically implied for the interface). Crucially, code that wants to call into a MyConcreteClass instance is allowed to refer to it as a MyInterface, just as a caller can treat any derived object as its base object.

When an updated version of the C++ language came out in 1989, it included the idea of multiple inheritance, which meant a class could inherit from two different, completely unrelated base classes. This was a clever idea in theory—and somewhere between neutral and terrible in practice. It turned out that the cases in which a derived class was 100 percent "is-a" with two different base classes rarely arose in any useful situations (a lot of the initial use of multiple inheritance in C++ was for inappropriate "has-a" relationships, such as inheriting from an encrypter class in an e-mail class). Multiple inheritance of classes eventually fell out of favor, and the two modern object-oriented languages that most people care about today, Java and C#, don't allow it. But they do support multiple inheritance of interfaces. This isn't in danger of violating the "is-a" rule because an interface "isn't a" thing at all; it's just a set of rules for what methods can be called.

Meyer, in discussing multiple inheritance, focuses on the value of interfaces (which he calls *deferred classes*, since Eiffel uses the keyword "deferred" for "abstract"); he calls it a "marriage of convenience," which sounds a bit negative until you read his explanation:

The FIXED_STACK example is representative of a common kind of multiple inheritance, which may be called the **marriage of convenience**. It is like a marriage uniting a rich family and a noble family. The bride, a deferred class, belongs to the aristocratic family of stacks; it brings prestigious functionality but no practical wealth—no implementation. The groom comes from a well-to-do bourgeois family, arrays, but needs some luster to match the efficiency of its implementation. The two make a perfect match.[3]

In my encrypting e-mail example, you would define an interface for encrypters, which looks similar to my definition of the AESEncrypter class above (interface names traditionally start with an *I*):

```
interface IEncrypter {
    Encrypt();
}
```

Any implementation of an encrypter can now inherit from the IEncrypter interface and implement the methods, as it normally would, except the methods now inherit their signature from the interface rather than having carte blanche to define them in each individual encrypter (this is a good thing: the enforcement of a standard method signature is the point of inheriting from the interface). You only need a one-line change in the declaration of a specific encrypter to add the inheritance from IEncrypter:

```
class AESEncrypter: IEncrypter {
    public Encrypt() { }
}
```

and then your e-mail program that needs encryption can use the IEncrypter interface to call through into whatever encrypter it is going to use, without your having to write extra code to decide which encrypter it is really calling into:

```
class Emailer {
    private IEncrypter e;    // can call e.Encrypt()
}
```

This leaves open the question of how e is initialized. In the worst case, you can have code in the constructor of Emailer (which is defined using the class-name-like-a-method-name syntax below) that sets it based on an encryption type. This still requires having a known list of encryption types and encrypters, but at least adding support for more encrypters only involves changing this one place, as opposed to every place you want to call into an IEncrypter:

```
Emailer() {
    if (encryptionType == AES) {
        e = new AESEncrypter();
    } else if (encryptionType == RC4) {
        e = new RC4Encrypter();
    } else if (encryptionType == TripleDES) {
        e = new TripleDESEncrypter();    }
}
```

But a much cleaner way is to have the constructor of Emailer take the IEncrypter as a parameter, saving it in e for later use:

```
Emailer(IEncrypter enc) {
    e = enc;
}
```

This arguably just pushes the problem up a level since it means the person creating the Emailer object now has to know how to create the IEncrypter so it can pass it to the Emailer constructor. There are a few solutions to that: you could write an Emailer constructor like the sample above that sets e based on an encryption type passed to the constructor but defaults to a specific encrypter if none is specified. In any case, hold that thought (I will come back to it later), and instead focus on how clean the internals of Emailer are with this design.

The concept that clarified the usefulness of interfaces was *design patterns*. Design patterns were first introduced in 1994 in a book of the same name, subtitled *Elements of Reusable Object-Oriented Software*.[4] The book was based on the 1977 book *A Pattern Language* by Christopher Alexander,

Sara Ishikawa, and Murray Silverstein.[5] That book has nothing to do with software; it is subtitled *Towns · Building · Construction* and is about architecture. The theme of Alexander and company's book is that there are design problems in architecture that have been solved repeatedly throughout the centuries, and it makes sense to collect, describe, and name them, so that architects can use and discuss them without lengthy reinvention or explanation. The architectural problems (there are 253 of them) range from large to small scale. Using the capitalization/numbering convention in the book as well as the book's general ordering from largest to smallest, they include AGRICULTURAL VALLEYS (4), RING ROADS (17), MARKET OF MANY SHOPS (46), BUS STOP (92), INDOOR SUNLIGHT (128), STREET WINDOWS (164), FLOOR-CEILING VAULTS (219), and HALF-INCH TRIM (240).[6]

The four authors of the software *Design Patterns* book (who became known as the Gang of Four) applied the same idea to common software problems. Like Alexander and his coauthors, they give each solution a name and describe it in detail. For example, the "Strategy" pattern addresses the problem we talked about above: a program wants to use a module to perform an operation such as encryption (the Strategy pattern is often demonstrated using an encryption module) in a clean way that makes it possible to change the encryption module or add more encryption modules with minimal changes to the code. The solution in the Strategy pattern is, as you might expect, to access the encryption module through an interface rather than inherit from or contain a concrete encrypter class.

The Gang of Four book is a bit dense, and some of the patterns are less insightful than others. For instance, the "Singleton" pattern consists of this code, which ensures that anybody who calls it will receive back the same instance of `SingletonClass`, which the code will create if it doesn't exist yet:

```
if (instance == null) {
    instance = new SingletonClass();
}
return instance;
```

There is nothing wrong with this, but it's more obvious than revolutionary; it's closer to the level of HALF-INCH TRIM (240). Meanwhile, the Strategy pattern can help you with a cleaner, more extensible design for plugging in encrypters, but it won't help you with the more common type of problem, which is that the person writing code to call an encrypter doesn't understand exactly how the `Encrypt()` method works and makes a mistake that cause a crash, or leaves the data unencrypted.

Still, it is useful to start creating a common language. "I will use the Singleton pattern" is shorter and clearer than "I will have a method that checks if the instance exists and allocates it if it doesn't." Singleton is the simplest pattern, although not by much: the patterns involve an arrangement of at most two or three classes. In his essay "The Evidence for Design Patterns," though, Walter Tichy makes the point that programmers, like anybody else, have a limited number of things they can keep in their short-term memory; using the names of patterns allows them to collapse a set of classes down to a single pattern, which leaves more room for other ideas.[7] And indeed the clearest benefits ascribed in Tichy's paper are related to documentation and communication of code details, not better design.

The Gang of Four stated two principles of good object-oriented design that were present in all the patterns: "favor object composition over class inheritance" and "program to an interface, not an implementation."[8] The first one means "don't do what we had in our early example, where `Emailer` inherited directly from `AESEncrypter`; instead contain an instance of the encrypter as one of your class data members." And the second one says, "That encrypter you are containing? Make it an interface like `IEncrypter`, not a concrete class like `AESEncrypter`." This gives you the maximum flexibility to change the internal details of the encrypter you are using or change your `Emailer` to use a different encrypter.

Design patterns are useful, but their anointing as the solution to all design problems has been imprudent. Erich Gamma, one of the Gang of Four, when discussing the misconception that more patterns always make a system better, later wrote, "Patterns make it easy to make a system more complex. They achieve flexibility and variability by introducing levels of indirection, which can complicate a design. It's better to start simple and evolve a design as needed."[9] There is a good idea at the core of design patterns, but the standardization is on the level of standardizing the sizes of nuts and bolts used in carpentry; it's a great benefit, although nobody would claim that it fully solves the problem of building a building that won't fall down. The Strategy pattern is elegant, but the real design questions—the ones that affect whether your software crashes, runs fast, or is taken over by Russian teenagers—are all far removed from it. For what it's worth, the patterns are presented as fact, without any studies to back up their benefits. Nonetheless, time has shown that the design patterns are clean solutions to the problems they address, as anybody who has spent time separating an improperly commingled "has-a" base and derived class can attest.

Plus, design patterns have an aspect that I really like: they were developed through a productive collaboration between industry and academia. The Gang of Four consisted of an IBM researcher, a Swiss programmer, a consultant, and a professor at the University of Illinois. Design patterns originated as the PhD thesis of one of the Gang of Four (the Swiss member, Gamma), and the idea was incubated at various OOPSLA conferences in the early 1990s. This is exactly how industry and academia should interact. Unfortunately, it's the only shining example from the past three decades. And the reason that design patterns could evolve as an industry/academia collaboration is precisely because they are not that complicated; their scope is at the level at which a professor can still engage and understand code.

There's another thing to note about patterns. Recall our discussion in chapter 3, concerning the mental anguish of a programmer deciding how much to plan for future changes versus solving the problem in front of them: a lot of the benefits of patterns relate to future extensibility. If your e-mail program uses a single encrypter, then the whole inheritance versus containment versus interface argument is somewhat pointless; the code you write won't differ much in complexity or size whether you follow the Strategy pattern or not (and if it does differ, it's the Strategy version that will be slightly more complex). It's when you want to modify it in the future that it matters, in terms of how much code you have to tweak or replace to make it easy. You need that "second thing"—a second type of encryption, say—for the pattern to pay off.

Luckily, at around the same time that patterns were emerging, another good idea was percolating in the world of programming, guaranteeing that you would need that second thing in all your code. This was the idea of the *unit test*.

The concept of unit tests, as used today, has a murky origin, since the term has been in circulation for a while, and the question is how big a unit you are talking about. The modern meaning was certainly present in a paper written by Kent Beck in 1989 titled "Simple Smalltalk Testing: With Patterns." It looks at testing a call to a single method on an object (since this is Smalltalk, it's actually about testing a single message to an object), which is much more localized than a lot of quote-unquote unit tests were at the time.[10] The idea of *automated testing*, writing a separate program whose only purpose was to test the program you shipped to customers, had been around for a while. Typically these automated tests would operate against the user interface of the software, simulating keystrokes and mouse clicks, and verifying that the program responded as expected. They were functionally the same as a human tester, but with the benefit of easy repeatability. The problem is the same one that occurs when humans do the testing: the user interface of the software is stacked on top of many layers of method calls, so if the software behaves incorrectly at the top level, it is hard to isolate the fault among all those layers.

Let's say your program has a method to sort an array of numbers, not directly accessible to the user but instead buried deep for use by intermediate code. In traditional user-interface-based testing, it may be hard to directly test this code. Still, you want to be confident that your sort routine works correctly. A unit test is code (test code that does not ship to customers) that directly calls the sort method, passing it an unsorted array, and then checks at the end that the array was sorted. The unit test doesn't randomly generate an unsorted array; it passes in a known array, for which the sorted version is also known, and after the sort is done, it compares the resulting array against the known sorted version.

Bugs in sort routines may manifest themselves only with certain inputs. Perhaps the routine fails to sort properly only if the first element in the array needs to move to a different location in the array. Thankfully, there is nothing stopping you from writing multiple unit tests for the same method; you can pass in an already-sorted array, an array that is in reverse order, an array where no element is in the correct place, an array with all the elements equal, or whatever other arrangements you think might ferret out a bug. The goal is to have lots of tests that run quickly: "A unit test that takes 1/10th of second to run is a slow unit test," as the unit testing advocate Michael Feathers put it.[11]

Some well-known bugs throughout history were clearly crying out for unit tests. Consider the bug that hit Microsoft's Zune media players on the last day of a leap year, as discussed in chapter 5. This is precisely the sort of thing a unit test could check. When testing the entire Zune, it might not occur to a tester to set the clock to the last day of a leap year, because they didn't realize that any unusual code ran on that day. But the developer who wrote the `ConvertDays()` function, aware that the code special cases the leap year, would presumably want to write a unit test that checked that the last day of a leap year was handled correctly.

Such a method is easy to test because it is self-contained, depending only on a value passed as a parameter; the code is all calculations based on that. What if the method you want to unit test is itself dependent on another method and so on through many layers? And worse, what if some of those methods have side effects that you don't want to occur during testing? You may be writing a unit test to verify code that formats data and then sends it to the printer. You can't have your unit test print anything, for a variety of reasons: it takes a while and wastes paper, and also requires you to have a printer connected and turned on, and even then how would your unit test verify that the proper printing happened within 1/10th of a second?

The cleanest approach is to write a fake printer object that supports the same methods as the real printer; you somehow tell the code that you are testing, which is using the printer, that it should use the fake printer instead. Without applying a design pattern, this might wind up with a bunch of code surrounding each call to the printer methods, like this:

```
if (runningTests) {
    fakePrinter.Print();
} else {
    realPrinter.Print();
}
```

which would need to exist every place you called any of the printer methods. If this reminds you of the prepatterns code above for selecting an encrypter to call, that's no coincidence. The Strategy pattern, the canonical design pattern, works well here, even if the name Strategy is a bit of a misnomer here; think of it as a pattern for "I want to take something that might change and encapsulate it behind an interface." You define an interface for the printer, which the real printer class inherits from and therefore implements, and also write a fake printer class that inherits from the same interface. The fake printer collects whatever it is told to "print," and includes extra methods that let you validate that what was sent looks correct (these methods exist only on the fake printer class, not the real one).

About ten years after the original *Design Patterns* book came out, Alan Shalloway and James Trott wrote a book called *Design Patterns Explained* in which they added a new rule that was implicit in the original design patterns: "Separate the code that uses an object from the code that creates an object."[12] This relates to the question that might still have you scratching your head. In the example above,

```
Emailer(IEncrypter enc) {
    e = enc;
}
```

we have the constructor of `Emailer` taking an `IEncrypter` interface as a parameter and saving it to use later in a nicely abstracted way, but where does that `IEncrypter` come from? The answer is to provide a separate class, known as a *factory*, which knows how to create a concrete encrypter class. Then any code that needs to construct an `Emailer` can call the encrypter factory to get the encrypter object and pass that to the `Emailer` constructor. This makes the creation of the real `Emailer` easy, while preserving the ability to construct an `Emailer` with a fake `IEncrypter` when running unit tests against `Emailer`.

You can do a similar thing in my printer example, allowing you to set up your unit test to use the fake printer with zero changes in the code you want to test. The unit test is the only code that is aware that the printer interface passed in to your code is in fact the fake printer, and is the only one that calls the extra verification methods that the fake printer object supports.

In this model, the fake objects (generally known as *mocks*) that are required for unit testing become the "second thing" that your code needs to call, so even if your real code is only going to ever use one printer class, unit testing will require it to use a second printer class, and therefore the convention that the design pattern dictates is not overkill but rather a simplification.

To be clear, unit tests are not perfect. Just because you write a unit test for a method doesn't mean it will work in all cases; you could write a unit test for a sort that handed it an already-sorted array, and the unit test would likely pass whether the algorithm was broken or not. One metric used to evaluate how complete your unit tests are is known as *code coverage*, which is the percent of your product code that is run if you execute all your unit tests. Ideally you would hit 100 percent, but even that won't guarantee correct code; the Zune bug was ultimately due to code that was completely missing (an ELSE branch of an IF statement), so a set of unit tests with 100 percent code coverage might still have missed the bug.

While they don't guarantee that your code has no defects, unit tests are a good way to guard against it getting worse. As we have seen, software is frequently expanded beyond its original purpose, and as part of this growth, code has to be reworked to accommodate the change. Some changes can be inserted with little risk, especially if it is the sort of change that design patterns allow—adding a new type of encrypter, for instance. But often the changes are more complicated, and the risk of accidentally breaking something is high. Having a good set of unit tests that continue to pass 100 percent is a great way to catch this quickly. If you recall the bug with doctors' addresses that was inadvertently introduced back when I worked at Dendrite, this is exactly the sort of thing that a unit test would likely have caught at the time the change was made, but because the feature that broke was not the one being added at that time, normal spot-checking by the developer didn't catch it.

Furthermore, unit tests can guard against your software getting stale. It is an unfortunate fact of software development that your program can start to fail without your having changed your code at all; a new version of another company's code library, an update to the compiler you use, or a patch to the operating system that it runs on—all these could break your code in an unpredictable way, through no fault of your own. This tends to prevent software from being moved forward, keeping it locked to older versions of whatever technology underpins it, requiring users to keep that older technology running. How many of you have a dentist who runs a version of Windows that is more than ten years old? Having a strong set of unit tests allows software to be moved to newer systems with much less fear of unintended consequences.

Unit tests are a great idea, and arguably anybody who is not writing them is not doing their job as a programmer. Certainly if you want to make software less fragile, unit tests are worthwhile, and if we are thinking about how to turn software into a real engineering discipline, not writing unit tests would be akin to building a bridge without first calculating if you expect it to stay standing. Requiring unit tests is a great way to emphasize to developers that they are responsible for the quality of their own code and get away from the "throw it over the wall to the testers" attitude that used to prevail.

The biggest hurdle to unit tests is that a lot of software was written without any consideration for them. If you already have a mock printer class written, and your code that uses the printer accesses it through an interface, and has properly separated creation and use, then writing one more unit test is fairly simple. If none of that support is there, and you need to write the entire mock printer class as well as go in and replace all the hard-coded printer calls with calls to an interface, and figure out a way to pass the interface through to the code that uses it, then writing that first unit test becomes a high mountain to climb, and you risk accidentally breaking your code while reworking it to support unit tests (a truly ironic result).

In 2004, Feathers published the book *Working Effectively with Legacy Code*—a title that speaks to one of the great frustrations of being a programmer: being handed the task of fixing a bug in a large piece of old software, where you don't know enough about the details (if you know anything at all) to ensure that fixing the bug won't break something else (another book on the same subject has the title *Software Exorcism*).[13] Although the phrase *legacy code* is often used to mean "old code that I don't understand," Feathers's book straight up defines it as "code without tests," and the book could be summed up as "before you touch anything, you need to have unit tests in place."[14] For example, you might fix the "last day of the leap year" bug in the Zune clock driver code, and of course you would verify that it worked correctly in that specific situation, but your change might inadvertently introduce a different bug—maybe it won't work on the first day of a leap year, last day of a regular year, or who knows—and if there are no existing unit tests, being the first person to decide to write a unit test means signing up to do a lot of work—potentially hundreds of times as much work as just fixing the bug and crossing your fingers. Given that programmers, as a species, introduced entire classes of security errors in C code just to avoid a little extra typing, it's understandable why people are unwilling to be the first person into the unit test breach.

There is something else at play here. Although the general idea of unit tests has been around for a while, the notion that they are a core part of what a programmer has to deliver has only gained strength in the last decade or so. A manager who didn't come of age with this idea will instinctively come up with estimates (of how long a piece of software will take to write) that are completely out of kilter with the reality of writing quality code. Splitting your time evenly between writing code and writing unit tests is a reasonable guideline, but this math will sound wrong to any old- or not-quite-so-old-school managers.

If the structured programming push wound up boiling down to "don't use GOTOs," that is still a worthwhile result. And if the main contribution of design patterns is making your code amenable to unit testing, that's also worthwhile: the unit testing message has tunneled past all the self-taught knowledge and embedded itself in the skulls of programmers, which is quite an accomplishment. Still, it does bring up a question: If design patterns and unit tests only help with small-scale design and testing, what can help with the broader design of big software systems? Once the personal computer software industry had, after much suffering, relearned the lesson that you could not "test in" quality, there was a push to "design in" quality instead. Who would provide the design expertise?

If you peruse job listings for programmers, you may see a job titled "Software Architect." In college, I heard somebody talk about software they had designed during a summer job (I think it was for a bank), and they said, dismissively, "At that point I handed it off to a coder." Whether or not summer interns get to hand anything off to anyone, this does capture the essential notion of the software architect: just as a "real" architect (one who works on buildings) comes up with designs and then hands them off to other people to build, a software architect would come up with software designs and then hand them off to others to write.

This is fine, but architects go to school to learn about architecture; they are not construction workers who have been doing it for a while and have a track record of nonfailure, which is how you become a software architect. Architects can justify their designs to a builder because they are based on an industry consensus on what works and what doesn't work, which in turn is based on a legacy of experiments and mathematics—the sort of thing that was captured in Alexander and crew's *A Pattern Language*. Software architects, like all programmers, rely instead on the "I tried it this way once and it wasn't too bad" approach. Only in the nuts-and-bolts areas covered by design patterns is there any equivalent of the built-up knowledge that an architect has, and a software architect would not stoop to the level of specifying details as nitty-gritty as how your code should call an encryption algorithm, any more than a building architect would tell a construction worker how to hold a nail gun.

Joel Spolsky is a noted software blogger who wrote a post about "Architecture Astronauts": software architects who like to think about higher and higher levels of abstraction. His description of an Architecture Astronaut at work is typical:

When great thinkers think about problems, they start to see patterns. They look at the problem of people sending each other word-processor files, and then they look at the problem of people sending each other spreadsheets, and they realize that there's a general pattern: sending files. That's one level of abstraction already. Then they go up one more level: people send files, but web browsers also "send" requests for web pages. And when you think about it, calling a method on an object is like sending a message to an object! It's the same thing again! Those are all sending operations, so our clever thinker invents a new, higher, broader abstraction called messaging, but now it's getting really vague and nobody really knows what they're talking about any more.[15]

Spolsky gives a warning sign: "That's one sure tip-off to the fact that you're being assaulted by an Architecture Astronaut: the incredible amount of bombast; the heroic, utopian grandiloquence; the boastfulness; the complete lack of reality. And people buy it! The business press goes wild!"[16]

The commonly prescribed antidote to software architects who only spout architecture is to require them to write product code that ships to customers, in order to keep them grounded and ensure their architecture is relevant. Whether this was a good use of their time was a subject of debate inside Microsoft, like many things; at one point I was involved in a series of discussions with people across the company to try to document "what makes a good architect," which foundered on disagreements such as this one. Nevertheless, the current bias is more toward the "sometimes-coding" architects than the "pontificating-only" ones. Although grounded software architects, at this moment in time, are considered better than oxygen-deprived ones, the fact that architects need continual immersion in their team's current project is another sign that there is not enough accepted knowledge and vocabulary around software engineering. Software architects *should* be able to leverage precedent to design a solution in the abstract, and be able to communicate that to any programming team in language that is clear and standardized enough that the team will recognize the value of the design, and be able to trust that it will be followed. Although people may complain about the designs of building architects, you never hear that the solution is having them occasionally hang sheet-rock just to ensure that their buildings will work.

Underlying the existence of the software architect role is the idea that there is "good" and "bad" design, and architects will choose the first and not the second. I'm talking about the underlying design of the software—the part that is not visible to the user (user interface design is a whole other area, outside the scope of this book). There is a sense that good design will show through to the user in some way, but I see no evidence that the user knows or cares about how anybody embeds their encryption algorithm in their code.

Beyond design patterns, what does good design look like? With a lack of theoretical rigor to underpin it, this is a murky area. One study of software design by Antony Tang, Aldeida Aleti, Janet Burge, and Hans van Vliet put it this way:

Software design has certain characteristics that are different from other engineering design disciplines. First, designers often have to explore new application and technology domains that they do not have previous experience with. Therefore, the quality of their design outcomes may not be consistent even for a designer who has practised for years. Secondly, a design is an abstract model and, often, whether it would actually work or not cannot be easily judged, objectively, until it is implemented.[17]

There are books that claim to explain good design—with titles like *Clean Code* and *The Pragmatic Programmer*—that are full of completely reasonable advice, but they don't present a specific approach to engineering your software.[18] They are more about lists of things to remember to do: don't forget to think about making your code localizable (meaning it can be translated into other languages), check frequently to make sure your code builds successfully, and the like.

The physician and writer Atul Gawande wrote a book called *The Checklist Manifesto* about how medicine can be made safer by using checklists.[19] One of his discoveries is that checklists don't need to be extremely specific to be helpful; for example, it is much more useful to have a checklist question like "Has the doctor discussed the anesthesia plan with the anesthesiologist?" than have a complicated checklist of all the steps the anesthesiologist should take. You don't need all the steps written down because the anesthesiologist has been to medical school and done advanced training in anesthesiology. Problems are more likely to arise from communication issues than from lack of medical knowledge. It would be great if we could adapt this approach to software, asking questions like "Has the developer discussed the test plan with the tester?" Yet there is not enough shared knowledge for this plan to work; what you see instead, in books that have software checklists, is long lists of specific things to worry about, making checklists hard to apply in practice.

In 1971, in *The Psychology of Computer Programming*, Gerald Weinberg wrote, "We shall be hampered by our inability to measure the goodness of programs on an absolute scale. But can we perhaps measure them on a relative scale—can we say that program A is better or worse than program B? Unfortunately, we will generally not even be able to do that, for several reasons. First of all, when is there ever another program with which to compare?"[20] This is an important point. Since programs are designed "from scratch" each time, it is always easy to see why a new program is slightly different from an existing one, in ways that make it invalid for comparison purposes. Of course, being unable to measure the goodness of programs makes it hard to measure the goodness of programmers, and in particular it makes it hard to measure *progress* in either of those areas, which you hope software engineering would be achieving after all these years.

For part of my career at Microsoft, I worked in a group with the slightly overambitious name Engineering Excellence, which did internal training and consulting. In response to demand, we created a course for software developers on how to design software. We came up with the title Practical Design for Developers, which we were quite proud of. Here was the class that was going to strip away all the nonsense and give developers the knowledge they really needed! And the class was quite well attended, so we weren't the only ones who thought that developers were hungry for this information. But after a while we realized that we were less and less confident that the class was useful; when you strip away the nonsense from software design, you are left with design patterns and not much else.

There is another truth about good design: it often runs counter to design that executes quickly. And given the performance focus in which many of today's programmers were steeped, it is hard to fight that.

Weinberg wrote about Fisher's fundamental theorem, derived by the statistician and biologist R. A. Fisher: "A word of caution before we proceed to the question of efficiency. Adaptability is not free. ... Fisher's fundamental theorem states—in terms appropriate to the present context—that the better adapted a system is to a particular environment, the less adaptable it is to new environments. By stretching our imagination a bit, we can see how this might apply to computer programs as well as to snails, fruit flies, and tortoises."[21] In other words, the more you optimized your program for speed, the harder it was to modify it later to accommodate extra functionality.

Computer scientists have long recognized that as computers have gotten faster, squeezing out every ounce of performance was no longer the primary goal. In *Structured Programming*, in 1972, Dijkstra wrote:

My conclusion is that it is becoming most urgent to stop to consider programming primarily as the minimization of a cost/performance ratio. We should recognise that already now programming is much more an intellectual challenge: the art of programming is the art of organising complexity, of mastering multitude and avoiding its bastard chaos as effectively as possible.

My refusal to regard efficiency considerations as the programmer's prime concern is not meant to imply that I disregard them. ... My point, however, is that we can only afford to optimise (whatever that may be) provided that the program remains sufficiently manageable.[22]

Recall Bentley's similar warning in *Writing Efficient Programs* that changes to make a program run faster "often decrease program clarity, modularity, and robustness."[23] Knuth phrased it this way in 1974, ending with one of his most famous quotes:

There is no doubt that the grail of efficiency leads to abuse. Programmers waste enormous amounts of time thinking about, or worrying about, the speed of noncritical parts of their programs, and these attempts at efficiency actually have a strong negative impact when debugging and maintenance are considered. We *should* forget about small efficiencies, say about 97% of the time: premature optimization is the root of all evil.[24]

Regrettably, this wisdom was lost on the new generation of programmers who taught themselves to program on personal computers, with strict resource limits that pushed programmers away from good design. They learned, on their own, the same bad pro-performance, antidesign wisdom that Dijkstra and Bentley had been fighting a generation earlier. This is not surprising for individual programmers: performance issues can be observed in a program of any size, and any improvements made can be measured, to positive effect on the psyche. Good design, on the other hand, matters when you are working on larger programs involving more people for longer periods of time—a situation that doesn't arise for programs within the scope of one person. Even old-fogy programmers, who arguably should know better, will focus their complaints about "those young kids" on a situation where they ignored performance rather than when they ignored good design; performance problems are more clear-cut and therefore easier to call out.

Some of the best-known problems in software were due to performance versus design trade-offs. The Y2K problem didn't arise because nobody realized that storing only the last two years of the date made 1900 indistinguishable from 2000. It arose because storing dates with two digits is slightly more efficient, and the immediate performance savings were viewed as worth it when balanced with the likelihood that the software would still be around when the year 2000 arrived.

Believe it or not, a similar situation is looming—the Year 2038 problem.[25] On many UNIX systems, times are stored as the number of seconds since January 1, 1970 (more precisely, since 00:00:00 on that date, or the stroke of the new year). Using a signed 32-bit number to hold the date, this will hit the maximum allowed value at 03:14:07 on January 19, 2038—exactly 2,147,483,647 seconds after January 1, 1970. Because of how signed numbers work, 1 second later will be interpreted as 2,147,483,648 seconds *before* January 1, 1970, or 20:45:52 on December 13, 1901—which, appropriately, is a Friday.[26] Again, this was not unknown to the designers of UNIX; they faced a trade-off of storing the dates using 64-bit numbers (or another accommodation that avoided the 2038 problem) versus the likelihood that UNIX would still be around in 2038. Back when UNIX was being invented, operating systems were less platform independent, and people tended to buy combinations of hardware and operating system together. It was reasonable to expect that when the next generation of hardware arrived, a new operating system would come with it, displacing UNIX. In that sense, those programmers were guilty mostly of underanticipating how well UNIX would hit the sweet spot of what programmers wanted and allow an independent software industry to emerge on top of it that helped ensure its continued presence. (And for what it's worth, if you are optimistic about the long-term survival of our species, many of the Y2K fixes involved replacing 2 digits with 4, so now we have a faintly looming Year 10,000 problem. But that will be someone else's to deal with.)

A nontechnical person may assume that simpler designs are also faster. Somehow this just seems right. There's less to do, correct? Yet it doesn't work out that way; having less code doesn't mean your software runs faster. In fact, it's frequently the opposite.

Consider a real-world example: an overnight shipping company that wants to offer delivery anywhere in the United States. It has a collection of offices near major airports all around the country, and each office has trucks and drivers that enable it to collect outgoing packages in time to deliver them to the airport by 8 p.m., and can deliver incoming packages the next day if they arrive at the airport by 8 a.m.

The simplest algorithm is to pick a central location in the United States that is a reasonable flight time from every airport—let's say no more than 4 hours away—and build a sorting facility there. Every day the local offices collect packages and bring them to the airport by 8 p.m. Planes leave each of the airports at 8 p.m. and arrive at the central sorting location by midnight. Between midnight and 4 a.m., all the packages are sorted and prepared to be loaded onto the same planes. The planes then fly back to their local airports, arriving by 8 a.m. The local offices deliver those packages during the day, collect new ones, and the process repeats.

From the perspective of a local office needing to know what to do with a package, the algorithm is dead simple:

```
void routePackage(package) {
    package.SendViaPlane(centralHub);
}
```

This abstracts away a lot of complexity. There are a lot of moving parts involved, such as lots of trucks and planes, not to mention people, and the sorting facility is probably quite sophisticated. Throw in tracking, payment, and all that, and you have a lot of layers underneath that call to SendViaPlane(). But the algorithm (since we're talking about algorithms here, we can for the moment ignore real issues on the ground) has a certain elegance: all the packages are loaded on a plane at 8 p.m. and then taken off a plane at 8 a.m. The sorting facility might be complicated, but the complication is all encapsulated within the one building, like a well-designed class encapsulates its internals—nobody else has to know about it, and it can change as long as the external interface is the same. If you decide that a particular city is better served by a different airport, you have to tell the people in the office in that city (so they know where to go to deliver and pick up packages), and the pilots who fly between there and the central location, but nobody else needs to know about it.

This algorithm will deliver every package to the right place, but it can be quite wasteful. A package I am sending to my next-door neighbor will go all the way to the central sorting location, spending hours on trucks and airplanes, and then taking the reverse route back. This would result in unnecessary cost, either for the delivery service or me. If the service charged me the full cost of this, I might complain or use a rival delivery service that could do it more cheaply. And if it charged me a low rate because the net distance wasn't far, then the service would be internally inefficient, flying my package around on its own dime.

This gives the delivery service an incentive to streamline its service. For example, it might tell the local office that if it saw a package whose destination was served by the same office to set it aside and mix it in with the incoming packages to be delivered the next morning, thus avoiding flying it to the central location and back. The algorithm then becomes:

```
void routePackage(package) {
    if package.Destination == thisOffice) {
        package.Store(thisOffice);
    } else {
        package.SendViaPlane(centralHub);
    }
}
```

This is an overall efficiency improvement, but it complicates the algorithm. The local offices have to check all the packages instead of just tossing them on a plane. This adds time and expertise to the work they have to do.

Now imagine that you decide to further optimize your delivery service by not flying all packages to the central location. You realize that there is enough traffic between Los Angeles and San Francisco, say, that it makes more sense to fly a plane directly between those two cities. And maybe it makes sense to drive a truck from New York to Boston rather than fly a plane, so you need to check for that option also:

```
void routePackage(package) {
    if package.Destination == thisOffice) {
        package.Store(thisOffice);
    } else {
        if (TruckAvailable(package.Destination)) {
            package.SendViaTruck(package.Destination);
        } else if PlaneAvailable(package.Destination)) {
            package.SendViaPlane(package.Destination);
```

```
    } else {
        package.SendViaPlane(centralHub);
    }
  }
}
```

Pretty soon you have a complicated algorithm for how each office should route packages. Looking at the code above, can you convince yourself that there are no cases in which it does nothing at all (which would presumably be bad)—that it will always wind up calling either `package.Store()`, `package.SendViaTruck()`, or `package.SendViaPlane()`? The way the code is written, it *will* always do something with each package, but it takes a bit of thinking (or a thorough set of unit tests) to be confident of that, making sure every IF has an ELSE, and that whatever path is taken through the code, the package winds up going somewhere. Furthermore, each office now needs to have knowledge of which trucks and planes are traveling where, and when, which makes it harder to change those details later.

I am not claiming that it is a mistake for an overnight delivery service to make these improvements. My point is that optimizing the performance of an algorithm generally makes it more complicated, not simpler. Programmers think that design and performance are correlated, such that better design runs faster. In reality, they are frequently inversely correlated: simpler, more elegant designs run slower, and you improve performance by complicating your design with special cases.

Why is this so? Software design is really the design of abstraction layers, and a design that is pleasing to the eye has nice, clean abstraction layers. The version of the shipping algorithm that has clean abstraction layers is the first one I came up with: every outgoing package gets put on a plane to the central sorting location, and every incoming package comes off a plane from the central sorting location. It's clean, it's simple, and anybody can understand it. The complications arise when you try to optimize it. And given the tendency of programmers to want to make things efficient, they tend to complicate things a lot. Whether the performance impacts that need to be mitigated are imagined in a programmer's head before any code is written or manifest in the real world after the software is deployed, the clean design rarely survives.

One of the most basic ways in which performance and clean design battle each other involves how errors are handled—which winds up being quite a story.

8 Your Favorite Language

Let us return to the saga of computer worms, last seen when Morris unleashed his on the fledgling Internet in 1988. As computers became more interconnected over the ensuing decades, the opportunities for worms to reproduce grew, especially with Windows becoming the dominant platform. A worm's infiltration path will usually only exist in one operating system; Windows was a juicy target for worm writers to concentrate their energies on.

The core internals of Windows, known as the *kernel* of the operating system, are written in the C programming language. C is uniquely supportive of the programmer errors that allow buffer overflows—the mechanism by which an exploit is injected into computers (both Linux and Apple's macOS, currently the two main competitors to Windows, also have kernels written in C). When I was an engineer on the Windows team at Microsoft in 1999, we were given training on buffer overflows; this was the first time I understood the potential risk of remote exploits and how easy it was to make a mistake in C that allowed one. We attempted to scrub the code by reading it carefully, but we didn't catch everything. In July 2001, a worm known as Code Red, which exploited a buffer overflow just as the Morris worm had done, attacked computers running Windows. In addition to bogging down the entire Internet by transmitting itself around repeatedly, the worm bombarded the White House website with messages, crashing it (once an exploit gains control, the damage is mostly limited by the imagination of the worm author).[1]

Worms are often reported to the company that owns the code before being widely released, usually with details on where the defect is (although attackers may not have access to the source code, while constructing the exploit they will become familiar enough with the compiled machine code that they can make a reasonably precise guess as to what the original source code looked like). The fix is typically simple, since most buffer overflows are due to errors in calculating the length of the buffer or the length of what is being copied into the buffer. Companies can issue a *patch*, which replaces the defective code with the fixed version. In the case of Code Red, the patch had been available from Microsoft for a month before Code Red struck, but unfortunately many users had not yet applied it.[2] Once enough machines were patched, everybody could breathe a sigh of relief—until September 2001, when a worm known as Nimda struck (once again, a patch had been available before the worm was released).

The next year was quiet. Jim Allchin, the group vice president in charge of Windows, was quoted as saying, "We have gone through all code and, in an automated way, found places where there could be buffer overflow, and those have been removed in Windows XP [which shipped in late 2001]."[3] Alas, it was not to be. The year 2003 featured both the Slammer and Blaster worms, 2004 brought Sasser, and 2005 was Zotob's moment in the sun.[4] Microsoft, to its credit, has continued to invest in tools to automatically detect exploitable code, especially the most obvious mistakes, where too much data is copied into a buffer allocated on the stack.[5]

Attackers have started targeting applications, such as Microsoft Office, by creating documents, spreadsheets, and presentations that are specially crafted to cause a buffer overflow in the application when they are loaded. These would be considered viruses, not worms, since the user has to open the file for the exploit to take effect, but they can do clever things to entice the user to do this. Typically the exploit code will e-mail the infected document as attachments to all the user's contacts, with subject lines such as "You HAVE to see this to believe it" (the result is, "A user opens an attachment they received from a friend … and what happens next will shock you!"). Worms and viruses are also becoming less immediately destructive; rather than propagate rapidly and cause havoc (which leads to quick detection and patching), it is better to hide in the background, renting out the exploited computer to other users for off-hours Bitcoin mining, website attacks, and the like. Despite my understanding of how remote worms could attack the operating system, the realization that viewing a virus-infected JPEG image could also cause your computer to be taken over—and that Microsoft Office was therefore another target for attacks—was another "Wait, what?" moment for me.

While none of these recent exploits involve code calling an API that was fundamentally unsafe, the way gets() was for the Morris worm, some of them do involve the mistake of trusting data received off the network, including the Heartbleed attack from 2014, which was due to a flaw in the implementation of the Secure Sockets Layer (SSL) protocol.[6]

SSL, which is used to encrypt web traffic, supports what's called a *heartbeat* packet, sent to verify that a connection is still active. The proper response to a heartbeat packet is to copy the data in the incoming packet and send it back. The problem is that the incoming packet, in addition to containing the data to echo back, also has a 2-byte field inside it that indicates how long that data is, and some SSL implementations trust that length field without checking it against the actual length of the packet (which would be returned by the API that the SSL code calls to receive the packet). Two bytes can hold a value up to 65,535, so if the incoming packet claims to have 65,535 bytes of data but it really has only 1 byte, then the defective code will try to copy 65,535 bytes into the response, which will consist of the correct 1 byte from the incoming packet, followed by 65,534 bytes of whatever is in the memory heap after that. Since this code is frequently running on web servers, that memory may well contain passwords, credit card numbers, and other readily identifiable data. Heartbleed does not take over the computer, but it still can leak sensitive information.

Some Windows exploits, however, were classic cases of good intentions getting stuck on the horns of programmers who didn't understand an API they were calling. The details require a little backstory.

Recall that when storing strings in memory you need to encode them, since ultimately the computer is storing numbers and there needs to be a standard for how numbers are interpreted as characters. Back in the UNIX/MS-DOS days, the most common encoding was ASCII, in which printable characters have values between 32 and 126. This allows room for lowercase letters, capital letters, numbers, and common punctuation symbols. One ASCII character occupied 1 byte; since a byte can hold a number up to 255, there was also an extended ASCII character set that included other useful symbols in the range from 128 to 255. That's still not enough room to hold all characters in all alphabets, so there were in fact many extended ASCII character sets, known as *code pages*, each with its own unique set of characters in the 128 to 255 range. The default code page, "Latin US," had certain common currency symbols, Greek letters, and a collection of letters with accents, tildes, cedillas, diereses, and so on.[7] It also featured a complete set of single- and double-line-drawing characters for programs to visually construct boxes on the screen using only characters, which was important for early IBM PCs that did not support graphics at all.[8] In addition, there were separate code pages for, among others, Arabic, Greek, Cyrillic, Hebrew, Portuguese, and Turkish (this last included our old friends the dotted capital *i* and undotted lowercase *ı*, at positions 152 and 141, respectively).[9]

Figure 8.1 is the default code page, with each row showing 10 characters.[10] Meanwhile, figure 8.2 is the Turkish code page; notice that they are identical from 32 to 127, and only have about 50 characters that differ in the 128 to 255 range.

Once a user configured their computer to use the right code page, text would display as expected on the screen. Or they could open a document written for a different code page and see amusing gibberish. On every code page, the original ASCII characters 32–127 remained unchanged, so English text displayed the same no matter what code page the computer was

32– 39				!	''	#	$	%	&	'
40– 49	(	)	*	+	,	–	.	/	0	1
50– 59	2	3	4	5	6	7	8	9	:	;
60– 69	<	=	>	?	@	A	B	C	D	E
70– 79	F	G	H	I	J	K	L	M	N	O
80– 89	P	Q	R	S	T	U	V	W	X	Y
90– 99	Z	[	\	]	^	_	`	a	b	c
100–109	d	e	f	g	h	i	j	k	l	m
110–119	n	o	p	q	r	s	t	u	v	w
120–129	x	y	z	{	|	}	~	⌂	Ç	ü
130–139	é	â	ä	à	å	ç	ê	ë	è	ï
140–149	î	ì	Ä	Å	É	æ	Æ	ô	ö	ò
150–159	û	ù	ÿ	Ö	Ü	¢	£	¥	₧	ƒ
160–169	á	í	ó	ú	ñ	Ñ	ª	º	¿	⌐
170–179	¬	½	¼	¡	«	»	░	▒	▓	│
180–189	┤	╡	╢	╖	╕	╣	║	╗	╝	╜
190–199	╛	┐	└	┴	┬	├	─	┼	╞	╟
200–209	╚	╔	╩	╦	╠	═	╬	╧	╨	╤
210–219	╥	╙	╘	╒	╓	╫	╪	┘	┌	█
220–229	▄	▌	▐	▀	α	β	Γ	π	Σ	σ
230–239	µ	τ	Φ	Θ	Ω	δ	∞	φ	ε	∩
240–249	≡	±	≥	≤	⌠	⌡	÷	≈	°	∙
250–255	·	√	ⁿ	²	■					

Figure 8.1
MS-DOS Latin US code page

	0	1	2	3	4	5	6	7	8	9
32– 39				!	"	#	$	%	&	'
40– 49	(	)	*	+	,	–	.	/	0	1
50– 59	2	3	4	5	6	7	8	9	:	;
60– 69	<	=	>	?	@	A	B	C	D	E
70– 79	F	G	H	I	J	K	L	M	N	O
80– 89	P	Q	R	S	T	U	V	W	X	Y
90– 99	Z	[	\	]	^	_	`	a	b	c
100–109	d	e	f	g	h	i	j	k	l	m
110–119	n	o	p	q	r	s	t	u	v	w
120–129	x	y	z	{	|	}	~	⌂	Ç	ü
130–139	é	â	ä	à	å	ç	ê	ë	è	ï
140–149	î	ı	Ä	Å	É	æ	Æ	ô	ö	ò
150–159	û	ù	İ	Ö	Ü	ø	£	Ø	Ş	ş
160–169	á	í	ó	ú	ñ	Ñ	Ğ	ğ	¿	®
170–179	¬	½	¼	¡	«	»	░	▒	▓	│
180–189	┤	Á	Â	À	©	╣	║	╗	╝	¢
190–199	¥	┐	└	┴	┬	├	─	┼	ã	Ã
200–209	╚	╔	╩	╦	╠	═	╬	¤	º	ª
210–219	Ê	Ë	È	ı	Í	Î	Ï	┘	┌	█
220–229	▄	¦	Ì	■	Ó	ß	Ô	Ò	õ	Õ
230–239	µ	×	ú	Û	Ù	ì	ÿ		°	¨
240–249	–	±	¾	¶	§	÷	¸			
250–255	·	¹	³	²	■					

Figure 8.2
MS-DOS Turkish code page

configured for.[11] Thus most programmers at Microsoft, working in English, were oblivious to the frustration of choosing the wrong code page.

When Microsoft Windows first came out in the mid-1980s, it included its own code pages covering the same basic ground but different in their exact encoding (for one thing, since Windows always ran in graphics mode, the line-drawing characters were no longer needed).[12] As before, you had to choose the right code page for your system if you wanted it to display the upper 128 characters that you wanted, and as before, English speakers were blissfully unaware of all this.

Code pages worked reasonably well but ran out of gas for ideographic alphabets, such as the Japanese kanji, which has thousands of characters. To solve this problem, the decision was made that Windows NT, whose first version came out in 1993, would switch to an encoding system called Unicode.

Unicode dispenses with code pages and stores all characters in two bytes, which allows 65,536 (almost) possible characters. It's actually slightly more complicated than that, to allow even more characters, yet not in a way that matters here.[13] Unicode's large character count allows the Chinese and Japanese ideographic alphabets (the most common subsets, anyway; it currently supports somewhat over 100,000 characters, including, in recent revisions, more than 2,500 emoji) to happily coexist with Latin alphabets in all their accented varieties, as well as Cyrillic, Hebrew, Korean, and so on.[14] The trade-off, of course, is that it uses 2 bytes of storage for every character, so strings occupy more memory and take longer to copy around, but (for once!) this was deemed a worthwhile choice.

Unicode was overall a great improvement, but it had one unfortunate side effect, which is that programmers, used to the idea that 1 character occupied 1 byte, had learned to think of the terms interchangeably.[15] Unicode opens up an avenue for programmers to make "characters versus bytes" math mistakes that can lead to trying to copy data into a buffer that can only hold half what the programmer expects—the exact sort of error that causes buffer overflows. Ten Unicode characters take up 20 bytes of memory, so if you try to copy 10 Unicode characters into a 10-byte buffer, you will overflow the buffer by 10 bytes. This is unexpected for programmers used to single-byte characters, for which 10 characters occupy 10 bytes.

When Unicode was added to Windows, the C API for string manipulation needed to be extended. Recall that a string in C is an array of 8-bit values, terminated by a 0; a Unicode string was defined as an array of 16-bit values, terminated by a 0—that is, a 16-bit 0. The API `strlen()` calculates the length of a single-byte string by scanning for a 0 byte; an API `wcslen()` was added that calculates the length of a Unicode string by scanning for a 16-bit 0 (that `wcs` at the beginning, which is short for *wide character string*, is a cameo appearance by a Hungarian prefix).

The question arises of what `wcslen()` should return: Should it be the number of bytes used or the number of Unicode characters? For `strlen()`, the single-byte version, they were equivalent, so that didn't offer any precedent. It was decided that `wcslen()` would return the number of characters, and indeed that is what it does. That's not wrong—it's the most logical choice[16]—but this is one more of those situations in which you have to know how an API works or you can make a mistake. There is nothing obvious about `wcslen()`—despite the Hungarian prefix!—to indicate "returns a byte count" versus "returns a character count."

When you allocate a buffer on the stack in C, the easiest way to figure out its size is via an API (which is not really an API, but think of it as one for simplicity) called sizeof(), which tells you how much memory a variable takes. Consider the following code (wchar_t is a single-wide character— that is, a 16-bit Unicode character, the Unicode equivalent of char in a single-byte strings; why the extra _t is there is not worth explaining):[17]

```
wchar_t my_buffer[10];
```

This allocates an array of 10 Unicode characters on the stack. In this situation, sizeof(my_buffer) will return 20: 10 characters time 2 bytes per character.

The problems arose when programmers wrote code trying to figure out if a Unicode string they were processing would fit in a stack buffer. In the days of single-byte characters, it was fine to mingle strlen() and sizeof(), like this (note the − 1 part; you still have to account for that final 0 character):[18]

```
char sb_buffer[10];
if (strlen(other_sb_buffer) <= (sizeof(sb_buffer) - 1)) {
    // other_sb_buffer will fit into sb_buffer
}
```

But in Unicode land, the following, which merely replaces all single-byte code with its Unicode equivalent, is wrong:

```
wchar_t wc_buffer[10];
if (wcslen(other_wc_buffer) <= (sizeof(wc_buffer) - 1)) {
    // other_wc_buffer will fit into wc_buffer—WRONG!
}
```

because sizeof(wc_buffer) is 20, so you may attempt to copy a 19-character string—that is, 19 Unicode characters, which occupy 38 bytes and will overflow wc_buffer.

Of course, programmers can avoid this easily. They just have to remember that sizeof() returns a value in bytes, and divide it appropriately, like this (and if reading this code gives you a headache trying to match up left and right parentheses, you are not alone; it's annoying, but messing them up can lead to hard-to-find bugs):

```
if (wcslen(other_wc_buffer) <=
    ((sizeof(wc_buffer) / sizeof(wchar_t)) - 1)) {
```

which would then correctly compare `wcslen()`'s output to 9 instead of 19.[19] Unfortunately when you are accustomed to not doing this with single-byte characters, it is easy to forget. And while not every oversight of this type led to an exploit—many of them, because of where they were, only caused a crash, and some, because the buffer on the stack was large enough to hold whatever strings it was called on to accommodate, never caused a problem—a few of them did. We used to joke that the value of typing those 16 characters,

```
/ sizeof(wchar_t)
```

as compared to the cost of some of the exploits, made it clear that programmers were underpaid.

It's not that software is doomed to have this sort of bug; it's that C allows it because of the way it handles strings, and combining that with Unicode, with the extra potential to mix up character and byte counts, makes it more likely. I should clarify that if you wanted to, you could write your own set of routines in C that handled all string manipulation in a safe way. You would define a structure, let's call it `safestring`, which could hold the characters in a string and also contained the length of the string, and you would then write a set of functions to create a `safestring`, manipulate a `safestring`, and extract the underlying string data from a `safestring` (to pass to APIs that did use raw C strings). And then you would sign a solemn pledge to always call those functions when your code did any operations on strings.

You could do this, but the resulting code would run more slowly and require more typing by programmers.[20] One of the main reasons people liked C was because you could handle raw string buffers quickly and tersely. But C, for all its myriad charms, is not a safe language for handling network messages directly; for that you want a language that makes it impossible for a worm to sneak in via a buffer overflow, a language where buffer copying is always checked against the actual length of the buffer, no matter what the programmer does.

What sort of language is safe?

Object-oriented languages have the power to take string length calculations out of the hands of individual programmers. In a 1998 update to C++, a new built-in class named `string` was added, which held a string (actually it was called `std::string`, but I'll ignore the `std::` part here). One of the features of `string` was that it *overloaded* the + operator to do string concatenation; you could write

```
c = d + e;
```

when c, d, and e were all of type string, and it would just work; C++ operator overloading allowed a new class, such as string, to provide its own + operator that worked in a way that made sense for that class.[21] The details are inside the implementation of the string class, where they had presumably been carefully verified to work, and every use of the string class can leverage that care (there is also a similar class wstring for handling Unicode strings). So it was like the C safestring we talked about above, except that once you declared a variable as a C++ string, you had no alternative but to use the functions provided, and the overloaded operators meant you could do it without extra typing.

The problem with this is that you may get memory allocation errors when performing these operations. More precisely, the code inside the string class, which handles all these details, may get memory allocation errors. How does this information propagate back so your code can react? In C, the memory allocations are obvious, since you call malloc() explicitly, but C++ is doing allocations under the covers here. In fact, even in the earliest days of C++, if memory could not be allocated for an instance of an object, then the constructor would return a value of 0 instead of the new object, but most code did not check for that any more than it tried to guard math operations against overflow.

To understand how a language can handle this cleanly and reliably, let's step back a bit and talk about how programs determine that an error has occurred.

Every time your code calls an API, there is a chance it will report back that it was unsuccessful. Code, by pure tonnage, is primarily dealing with things going wrong, despite the fact that nothing goes wrong the vast majority of the time that it runs.

Cast your mind back to 1996 and imagine that you are a programmer writing code in C to run on Microsoft Windows. The little bit of code you have been tasked with today should create a file, write out 1,000 bytes to it, and close it. The result might look like the snippet that follows. The first line of code is truncated, since CreateFile() actually takes six more parameters besides the filename that specify precise details on how the file should be created; a typical set of values for these parameters would be, in order, GENERIC_WRITE, 0, NULL, CREATE_NEW, FILE_ATTRIBUTE_NORMAL, and NULL, but removing them makes it easier to read:[22]

```
handle = CreateFile("foo.txt");
WriteFile(handle, buffer, 1000, &written, NULL);
CloseHandle(handle);
```

Each of the three steps invokes an API with certain parameters: first create a file with the name foo.txt, then write the data to the file, and then close the file. We can draw an analogy with storing a piece of paper in a filing cabinet: first open the cabinet, then insert the piece of paper, and then close the filing cabinet. The use of the word *file* in the computer context is not an accident; it was meant to guide users toward the same analogy, although at this point it is probably more likely that somebody would explain a filing cabinet in terms of computer storage rather than the other way around.

You will notice that the variable handle appears in all three lines; this is a value that CreateFile() returns back to the program, which then becomes a parameter to WriteFile() and CloseHandle(), so that they can perform their operations on the correct file.[23] This is typical of the way that non-object-oriented languages impose a smattering of object-orientedish loose coupling, because handle is an opaque value the callers pass around; the code inside WriteFile() and CloseHandle() knows how to interpret handle to get more details about the file (often it is really a pointer to an internal data structure), but those implementation details can change without affecting calling code.

This code works fine if nothing goes wrong, and usually nothing will go wrong; the file will be successfully created with 1,000 bytes of data. Yet things could go wrong. For instance, the disk could run out of space during this operation; it could have, say, 500 bytes of free space before you started and therefore not have room to write the 1,000 bytes. This is extremely unlikely in these days of multiterabyte hard drives, but it could happen.

The most likely failure here is that a file by that name already exists. CreateFile(), despite what the API name might imply, can also open an existing file if it finds one with the requested name. In this case, since we specified CREATE_NEW (one of those extra parameters that I chopped out of the actual code listing, which means it "Creates a new file, only if it does not already exist"[24]), it will fail if the file already exists.

The three lines of code shown above completely ignore all these possibilities; the call to WriteFile() assumes that handle is valid, but this will only be true if CreateFile() succeeded, and so on. Nonetheless, this is perfectly legal code, and you may be using a computer application that is written this way; you won't know until you hit the unusual error case at runtime and the program doesn't deal with it correctly. Imagine if the code had a fourth line added, with the first three lines remaining unchanged (as with `CreateFile()`, I've simplified the call to `MessageBox()` by removing irrelevant parameters):

```
handle = CreateFile("foo.txt");
WriteFile(handle, buffer, 1000, &written, NULL);
CloseHandle(handle);
MessageBox("File written OK", MB_ICONINFORMATION);
```

That last line (calling `MessageBox()`) tells the system to display a pop-up window; the `MessageBox()` API is the C equivalent of the `Message-Box.Show()` method in C#. The message will be "File written OK," and the icon displayed will be the information symbol (in current versions of Windows, this is a lowercase *i* in a blue circle).

If you see this pop-up, you would naturally assume that the file was written successfully. After all, the message box told you so! But if the code looked like the sample above, that pop-up would guarantee nothing of the sort. It's possible that the call to `CreateFile()` failed, which would have caused the `WriteFile()` and `CloseHandle()` to fail also, so nothing would have been written to disk. Yet the code would have blithely displayed the message box.

This is not acceptable from a software engineering point of view, even if the compiler allows this sort of plonky code. A conscientious programmer can add code to warn the user if something goes wrong:

```
handle = CreateFile("foo.txt");
if (handle == INVALID_HANDLE_VALUE) {
    MessageBox("Couldn't open file", MB_ICONERROR);
} else {
    b = WriteFile(handle, buffer, 1000, &written, NULL);
    if (b == FALSE) {
        MessageBox("Couldn't write to file",
        MB_ICONERROR);
        CloseHandle(handle);
    } else {
```

```
        b = CloseHandle(handle);
        if (b == FALSE) {
           MessageBox("Couldn't close file",
           MB_ICONERROR);
        } else {
           MessageBox("File written OK",
           MB_ICONINFORMATION);
        }
    }
}
```

This code is more complicated because of the IF and ELSE statements; recall that C syntax rules state that if the test in parentheses after the IF is true, it executes the code between the first pair of braces (the { and } characters); if the IF is false, then it executes the code between the braces following the word ELSE (recall also that in the text, I am capitalizing language keywords like IF and ELSE, but I'm referring to the keywords that appear in the code above as if and else). The construction handle == INVALID_HANDLE_VALUE is interpreted as "is handle equal to the value INVALID_HANDLE_VALUE," with INVALID_HANDLE_VALUE being the value that CreateFile() returns if it fails. The other two APIs, WriteFile() and CloseHandle(), return a Boolean, a true/false value, to indicate if they succeed or fail (they will always fail if the handle passed in is INVALID_HANDLE_VALUE).

This code avoids the problem of telling the user that the operation succeeded when it actually failed. But instead of four lines of code, you have thirteen or possibly sixteen lines of code (the philosophical question "Does a line containing only a curly brace count as a line of code?" is another religious query bandied about among programmers). And still this is only a slight improvement; if something goes wrong, the pop-up serves as a warning, but doesn't help with recovery. The user doesn't have the data written to disk and may have lost work. To fix this properly, you would need to add even more code, perhaps tell the user what happened, give them a chance to try again, and so forth, thereby adding to the percentage of your code that is involved in error recovery rather than the mainline work. You wind up with twenty-five to thirty lines of code, with the actual functionality being still just the original four lines of code—a terrible signal-to-noise ratio.

In addition, the error-checking code is mixed right in with the main logic, making it hard to follow. Error-checking code is like an annoying kid watching you attempt something, repeating, "It's not going to work!" over and over, until one day it doesn't work, and then they can say, "I told you so."

There's another thing too, which seems trivial. As you move through your code and call more APIs that need to be checked for errors, each check involves an IF statement, with the code in the IF and ELSE blocks usually indented one more level for readability; you can see this in the sample above that calls three APIs. With even more API calls, your code would be even more indented, until visually it risks bumping into the right edge of your editor. In any modern language, you are free to split a single line of code up into multiple lines, but each resulting fragment of the line will also likely be indented just as much, and there is something about typing all those indents and lining up the line fragments that feels like unnecessary work—with the programmer desire to avoid typing springing up again. And mistakes lining up indented code can lead to real defects, if you don't do it correctly and associate the wrong block of code with a certain IF or ELSE case.[25]

This overall problem has been known in programming circles for a long time, and a solution has been proposed for a long time as well: write programs that depend on *exceptions* instead of *errors*.

The error approach is what we see in the code above: code that calls an API will immediately check whether the API failed, by whatever mechanism the API has documented that it will indicate failure—which could mean returning an invalid handle, returning a Boolean false, returning a specific error code, or something else, with no consistency between APIs. To paraphrase Tolstoy, successful API calls are all alike, but every failed API call fails in its own way. Proper detection is dependent on checking for errors correctly, through all the layers of software, many of whose code you can't see. Beyond ignoring errors, code may check for the wrong ones; if the caller of an API is checking for ERROR_ACCESS_DENIED but instead gets ERROR_INVALID_ACCESS, it won't catch it. There are a lot of these similar-sounding errors; in Windows, you have ERROR_FILE_NOT_FOUND and ERROR_PATH_NOT_FOUND, ERROR_WRITE_PROTECT and ERROR_WRITE_FAULT, ERROR_INVALID_FUNCTION and ERROR_NOT_SUPPORTED, and so on. And of course there is the wonderful ERROR_ARENA_TRASHED, which persists with horsetail-like tenacity as error #7 on a system where the errors were essentially handed out in the order they were needed when implementing it (the "system" being DOS 1.00, in this case, since those low-value error codes were defined in that era).[26] Code can attempt to make a nonspecific check for "any error at all," but that reduces the chance that it will do something clever to recover from a specific error, and instead you wind up failing back up through the chain of callers until the user gets a mysterious error message.

The solution proposed for all this, getting away from errors entirely, is to use what are known as exceptions.

Exceptions are like an on-call telephone list for a crisis at work: the employee responsible gets called, and if they don't answer, then their boss gets called, and so on up the organizational chart. Code that determines that an actual error has happened *throws* (in the vernacular of exceptions) an exception. The code that called that code can indicate if it knows how to deal with that exception; if it does (known as *catching* the exception), then it provides code to be run in this situation (for example, display a dialogue box to the user). If the code that called the throwing code does not catch the exception, then the code that called *that* code has a chance to catch it, and so forth up the chain.

Exceptions work even better with objects, because object-oriented languages define that when an object goes *out of scope*—meaning that the method it is declared in returns or the block of code it is declared in ends—it is automatically cleaned up by calling a special class method called a *destructor*, which is provided for this purpose. If the destructor is properly written, then the "cleanup on error" part of error handling will be taken care of as well, without the need for extra code in the method that uses the object. As shown below, the object-oriented equivalent of `CreateFile()` returns an object, not a handle, and when the object goes out of scope, it automatically closes the file if needed (at least it will if the implementation of its destructor, which you likely can't see the code for, is correct—fingers crossed). Exceptions are handled in an orderly way; as the exception handler walks up the chain of callers looking for one to handle the exception, objects will still get cleaned up as the exception propagates past the layer at which they were declared.

The code to catch an exception (I'll switch to C# here) looks like this, using the keyword `TRY` to indicate the code that will be checked for exceptions (if this same code is run outside a TRY block, then any exceptions would automatically propagate back to the caller of this code). The main logic is clearly laid out, with no error handling mixed in:

```
try {
    using (FileStream fs = File.Open("foo.txt",
            CreateNew)) {
        fs.Write(buffer, 0, 1000);
    }
} catch (Exception e) {
    // handle any exceptions here
}
```

The FileStream object returned by File.Open() is the equivalent of the handle returned by CreateFile() in the C illustration. The USING syntax is needed to ensure the FileStream will get cleaned up properly if an exception occurs; don't worry about that and instead focus on the fact that with one CATCH, you can handle any exception that happens anywhere in the chain of calls below this point.[27]

To a programmer raised on error checking, relying on exceptions can appear dangerous; previous programming experience, the usual guide, fails them here. Error checking reminds you of a room of people in dinner jackets smelling their brandy snifters, discussing the latest reports from the CreateFile() front; exceptions feel more like the controlled explosions used for avalanche prevention. Is everybody sure the explosion can be controlled? But this is excessive anthropomorphizing of code; exceptions can be reliably caught if the language supports it.

There is also an argument that having the destructor do cleanup (when it is automatically called when the object goes out of scope) hides the details of the cleanup, leading to potential bugs. Better to have the code that uses the object do the cleanup explicitly, so the code can be clearly seen. The problem is that having everybody write their own cleanup code, besides clogging up every program with duplicated cleanup logic, gives everyone the opportunity to code it incorrectly. Certainly the destructor is an API that needs to be well documented and understood by users of the class, which today is often not the case, but better to document its behavior than document the expected behavior that you need all callers to reimplement themselves.

The exceptions approach is not perfect. There is no guarantee that the code that runs after an exception is caught (the code in the CATCH block) will do the right thing; the most extreme example of this would be code that caught all exceptions but then did nothing in the exception handler, which is effectively the same as ignoring an error code (that is what the code above does as written, since all it currently has in the CATCH block is a comment saying "handle any exceptions here"). But you have to go out of your way to do the wrong thing, whereas with errors you have to go out of your way to do the right thing; if you write your code assuming that every API succeeds, in error-based code you will silently miss all errors, and in exception-based code you will crash on all errors (since if an exception propagates all the way back to the top level of a program without being caught, the program will crash). Exceptions steer programmers away from the situation where careless programming leads to the outcome, "We reported success to the user but there was actually an error."

We can now get back, finally, to our problem of a string class being able to reliably report memory allocation errors, which would allow you to use it with confidence when manipulating string buffers. With exceptions, the code inside the string class can throw an exception if it can't allocate memory. If code that uses the string class is oblivious to this possibility, then the exception will propagate upward rather than being cast aside the way it would be if the code had to explicitly check for an error right at the point it might happen. This makes it OK for code to hand off all the string-processing details—including memory allocation—to the string class, and we can avoid the risk of buffer overflows from reimplementing the string processing every time, as is done in C.

Another example is integer overflow, which has historically been ignored by programmers. In an error-based system, you have the same problem as detecting memory allocation errors while concatenating strings; if the code doesn't check for it right when it could happen, any errors slip by unnoticed. With exceptions, the code to do the integer calculations can throw an exception when it detects an overflow, so they can't be silently missed. C++ does not support this, but C# does, either in a single block of code or for a whole program, although somewhat disappointingly, this is turned off by default, presumably as a concession to programmers with a performance bee stuck in their coding bonnets. Microsoft offers the weak defense, "Because checking for overflow takes time, the use of unchecked code in situations where there is no danger of overflow might improve performance" (to be fair, enabling this by default could also expose non-failure-causing overflows in existing code, where ignoring an overflow manages to work somehow, but wouldn't it be better to uncover those sooner rather than later?).[28]

While it was a big step forward for C++ to support an exception-throwing `string` class in 1998, the language, for unavoidable historical reasons, is stuck in a middle ground. The errors versus exceptions debate is long running in programming circles; the first version of C++ did not support exceptions at all. Stroustrup considered it, but didn't feel he had time to come up with a good design. Exceptions were also not present in version 2.0, which added multiple inheritance and various other features. In 1990, Stroustrup, along with Margaret Ellis, published *The Annotated C++ Reference Manual*, which laid out C++ as he wished it to be, with exceptions included; it took a few more years for C++ compilers to catch up and support them.[29]

By the time the `string` class was standardized in 1998, exceptions were part of the language, so `string` could be defined to throw exceptions when memory allocations failed. In the years before `string` was standardized, however, people hand rolled their own equivalents so that they could take advantage of nifty features like overloading + for string concatenation. Modules using two different homemade string implementations might also differ in how they handled errors, making it difficult to knit them together. And C++ continued to support the old "0-terminated `char` array" style of strings (known as *C-strings* in C++). Stroustrup later reflected, looking back on the history of C++, "To my mind there really is only one contender for the title of 'worst mistake.' Release 1.0 and my first edition should have been delayed until a larger library including ... a simple string class could have been included. The absence of those led to everybody reinventing the wheel and to unnecessary diversity in the most fundamental classes."[30]

It is crucial to emphasize the uphill battle that Stroustrup faced in order to make C++ appealing to performance geeks for whom C was the one true language (while simultaneously dealing with object-oriented purists who thought any C-ness was an unacceptable compromise).[31] Exceptions are not a new concept; in the 1960s, PL/I had the ability to specify code blocks preceded by "ON conditions" (like ON OVERFLOW or ON ZERODIVIDE), which would be invoked if specific exceptional conditions were detected, and the basics of structured exception handling were laid out in papers by John Goodenough in ACM conference proceedings and publications in the mid-1970s.[32] But exceptions were always haunted by the notion that they were slow and memory hogging, and in particular that the overhead necessary to track how to handle things *just in case* an exception hit was always going to exist even if nothing went wrong, whereas code to check for errors would lurk harmlessly until it was needed.

Stroustrup had to go to heroic lengths to ensure that C++ did not have any unnecessary slowdowns; he worried about adding a few extra machine language instructions to the cost of calling a function as the price for supporting multiple inheritance.[33] The fact that he succeeded in getting C programmers interested in C++ is barely short of miraculous. Similarly to Caesar's wife, he had to avoid any appearance of unnecessary slowdowns, and having an exception model from the start—where the standard APIs themselves used exceptions, as they do in C#, thus forcing everybody to use exceptions—would have been a huge source of dissatisfaction among his target audience.

C++ is now defined by a standards committee, which publishes updates every three to five years, and exceptions are completely integrated into modern versions of the language. But there is still a lot of C++ code being maintained that predates exceptions, and more important, there are a lot of C++ programmers out there whose experience learning the language predates the wide availability of exceptions. To this day, there is considerable debate about whether using exceptions in C++ is a good idea or not; the arguments against it include the reputed runtime overhead (which is claimed by proponents of exceptions to be wildly overstated) and the fact that there's an existing codebase that doesn't use exceptions and would be a headache to retrofit (which is similar to the assertion occasionally deployed against unit testing). The result is that C++, while a critical link in the chain of languages, winds up being neither fish nor fowl but halfway between errors and exceptions. When I left Microsoft in early 2017, this debate was still ongoing even among teams within Microsoft Office, leading to a few gymnastics required for code that needed to run in both an exception-embracing and exception-shunning application.

Stroustrup once said, "There are only two kinds of languages: the ones people complain about and the ones nobody uses."[34] Certainly nobody would call C++ a language that nobody uses, so you can guess what comes next. Peter Seibel's 2009 book *Coders at Work* features interviews with fifteen software luminaries, and their opinion of C++ is not positive. Quotes include "C++ is hairy" (Brandon Eich), "I couldn't exactly bring myself to use C++" and "C++ was pushed well beyond its complexity threshold" (Joshua Bloch), and "I can hardly read or write it. I don't like C++; it doesn't feel right. It's just complicated" (Joe Armstrong).[35] And those are the milder comments. The people who really don't like it say things like "C++ is just an abomination. Everything is wrong with it in every way" (Jamie Zawinski), "the syntax is terrible and totally inconsistent" (Brad Fitzpatrick), and "it certainly has its good points, but by and large I think it's a bad language. It does a lot of things half well and it's just a garbage heap of ideas that are mutually exclusive" (Ken Thompson).[36]

Harrumph, as they say. Stroustrup has nothing to be ashamed of; what he accomplished, given the era in which he accomplished it, was remarkable. C++ was not the first object-oriented language, but it was the one that popularized the approach, which allowed design patterns and unit testing to go mainstream. Some of the people quoted in the book, while claiming they would not deign to line their birdcages with C++, are leveraging his spadework to write better object-oriented languages.

This brings us to a vexing problem. Programmers have access to more languages today than ever before, but there is not a lot of guidance on when to choose one language over another. As a result, they tend to continue using languages they have used before, even in situations where the language is far from the best choice. Beyond areas I have already covered, such as susceptibility to buffer overflows, languages vary in how easily they handle certain kinds of programming challenges. The corollary is that they vary in how prone they are to bugs when handling those same programming challenges. And despite what is implied by the term *language*, it is much easier to learn a new programming language than it is to learn a new spoken language. For programs at the scope that industry generally tackles, the long-term negative of language unsuitability quickly outweighs the short-term benefit of language familiarity.

In the old days, a lot of new languages were created, but they were often tied to one operating system; the corollary of this was that languages, even those with wide adoption at one time, would become obsolete, especially as the systems that they ran on became obsolete. Despite the large amount of code written in them back in the day, few people are voluntarily programming in Fortran (from 1957), Algol (1960), COBOL (1962), or PL/I (1965), except possibly to maintain old systems that managed to avoid getting defenestrated in the Y2K cleansing. By the time I was in college, twenty to thirty years after the introduction of those languages, they were already in clear decline. In 1986, I had a summer job modifying a Fortran program, which was already understood to be outdated; the urge to rewrite it in C was great.

More recent languages have tended to stick around; C is still popular as it nears its golden anniversary, as are C++ thirty years out, and Java, which originated in the early 1990s. Related is the fact that the most popular operating systems, Windows and various UNIX variants, have also stuck around for an unexpectedly long time, so old programs, including compilers and related tools, can continue to run for a long time. MS-DOS first appeared in 1981, but was considered obsolete fifteen years later once Windows 95 shipped; development of the current Windows kernel began in 1988, and there is no replacement on the horizon (the underpinnings of Linux and macOS also date from around the same time). The effect is that once programmers learn a language and become proficient in it, there are no longer any obstacles preventing them from continuing to use it for the rest of their career, no matter how unsuitable it may be for the job. The code that the Heartbleed worm attacked was written in C, which as I have discussed is a dangerous choice for that sort of code. Even in the late 1990s, when work first began on the implementation of SSL that was attacked, C could have been seen as risky, but certainly from today's perspective we know it is. But with C (and UNIX) still alive and kicking, there was no push to replace it, and programmers continue to work on the code, including, in 2012, accidentally adding the buggy code that enabled Heartbleed.[37]

If you look at the course catalogs of computer science departments, they generally don't devote time to comparing languages and choosing the right one for a task.[38] The ACM and IEEE Computer Society periodically come out with curriculum recommendations for universities, most recently in 2013.[39] This does a thorough job of laying out the various areas that a computer science curriculum could cover, but in the programming languages section, most of the courses are aimed at people writing compilers (an interesting and fairly well-understood area, mature enough that I myself took a university class on writing compilers in the 1980s, but of little value unless you wind up writing one yourself) rather than providing specifics about the pros and cons of different languages for solving different types of programming problems.

The report also lists the details of a variety of exemplar courses in different areas, and perusing these you do see notable exceptions. Carnegie Mellon has a Principles of Programming Languages course, taught by Robert Harper and based on his book *Practical Foundations for Programming Languages*.[40] The course has a summary in the report (written by Harper), which I will quote from since it speaks precisely to my point:

Programming language design is often regarded as largely, or even entirely, a matter of opinion, with few, if any, organizing principles, and no generally accepted facts. Dozens of languages are in everyday use in research laboratories and in industry, each with its adherents and detractors. The relative merits of languages are debated endlessly, but always, it seems, with an inconclusive outcome. Some would even suggest that all languages are equivalent, the only difference being a matter of personal taste. Yet it is obvious that programming languages do matter!

Yet can we really say that Java is "better" (or "worse") than C++? ... Can we hope to give substance to any of these questions? Or should we simply reserve them for late night bull sessions over a glass of beer? ... Programming language theory liberates us from the tar pit of personal opinion, and elevates us to the level of respectable scientific discourse.[41]

One effect of the course is turning programmers into discriminating consumers of programming languages; Harper told me that one of his students reported back, after working in industry for a while, "My coworkers know so little programming language theory that I can't even explain why Python [a popular programming language] is an abomination."[42] Unfortunately this course stands out as unusual among all the other ones described in the curriculum recommendations; while a few of them do get into the area of comparing different languages (and learning fundamental concepts so that a new language can be picked up more quickly), certainly none of the other descriptions are written with such urgency.

The net effect is the same as in other areas of software. Self-taught programmers assume that what they know is good enough and don't try to learn anything else. If they have a favorite language based on past experience, they will continue to use it because they can see no reason to change. Although C++ certainly has problems, in the anti-C++ quotes above there was an undercurrent of "I don't like C++ because it's different from my favorite language." C++ may have enough flaws that it is rarely the best choice, yet that same attitude is aimed at newer languages that may indeed allow for easier coding, with fewer bugs, than the language that the programmer was planning to use.

Rob Pike, one of the original UNIX crew, gave a talk on "The Unix Legacy" in which he described C as the "desert island language"—meaning the one language you would choose if you could only choose one for all your programming tasks.[43] C may well be the most general-purpose language available today, with the broadest availability of compilers and platform APIs of any language. But that doesn't mean I want to use it every day; I'm sure if I were stuck on a desert island, I would eat whatever I could get my hands on, but back home I prefer a more varied diet. Pike said that "C is well understood and has aged surprisingly gracefully"; it is indeed impressive that a language that old still has any relevance at all. Yet the industry needs to understand that C attained its permanence from a unique event, when self-taught programmers obsessed with performance went off to college and discovered it, and then never reevaluated their opinion.

"Programming style is what results from writing programs under a set of constraints," notes Cristina Lopes in her book *Exercises in Programming Style*: "Constraints can come from external sources or they can be self imposed; they can capture true challenges of the environment or they can be artificial; they can come from past experiences and measurable data or they can come from personal preferences."[44]

Lopes's book, in a manner similar to how Alexander and company's book on architectural patterns led to the Gang of Four's book on design patterns, was inspired by the French writer Raymond Queneau's book *Exercises in Style*, in which he wrote the same short story ninety-nine different ways (some of them admittedly a little unreadable, such as permuting the letters according to a pattern or ordering the words by part of speech).[45] Lopes shows the same short programming problem solved thirty-three ways, demonstrating how different constraints affect program design—not just the usual constraints such as minimizing memory usage or program length, but issues such as use of objects or how deeply the code is layered. Students, especially those self-taught in high school, may be unaware of how the constraints of their early programming environments affect the style they use even when those constraints no longer apply. As my former professor Henry Baird put it, "It is hard to teach someone that they need to eat a thousand meals before they cook one."[46]

Harper, in his pitch for programming language theory in the ACM/IEEE curriculum recommendation, continues:

"Little languages" arise frequently in software systems—command languages, scripting languages, configuration files, mark-up languages, and so on. All too often the basic principles of programming languages are neglected in their design, with all too familiar results. After all, the argument goes, these are "just" scripting languages, or "just" mark-up languages, why bother too much about them? One reason is that what starts out as "just" an ad hoc little language often grows into much more than that.[47]

The *scripting languages* that he mentions are heavily used these days. They are descendants of the language used in the UNIX command line environment, usually focused on manipulating strings easily; the original use was for throwing together quick programs for small operations, such as renaming files in a particular way. When writing scripts like this, it is useful to smooth over differences in the data. For example, Perl, one of the best-known scripting languages, considers the string "0" and the number 0 to be equal, which is not true in most languages; you have to explicitly convert the string to a number first. Similarly, in Perl if you look up a value in an array using an invalid index, it won't crash like most languages will, but will instead return a special value undef. These are conveniences; they avoid writing a line or two of extra code—to convert the string to a number or check first if your array index is valid—which as always programmers are keen to avoid.

This is fine and works well for small scripts where you can easily verify if the program is doing what you expected, but as you start to write longer Perl programs, these clever ideas can cause bugs. If you write code to check an element in an array (something like this)

```
if ($arr[$index]) {
```

the IF will be false if there is no $index element in $arr, but it will also be false if there is such an element and its value is the number 0—and it will even be false if there is such an element and its value is the string "0." This can all be worked around and mitigated if you are careful, yet it can cause tricky bugs when it misbehaves in the guts of a large program just because a string happens to contain the value "0"; at some point the cleverness isn't so clever anymore.

And fans of Perl certainly do write long Perl programs; the language has grown over time to be a complete general-purpose language, including support for objects. One author called it the "Swiss Army chainsaw of scripting languages," which fairly accurately captures the power and danger of the language, and also the fact that it might not be suitable for all situations.[48]

Universities generally remain silent on this issue of language choice. If they expose students to Perl at all, it will be for smaller programs that are typical of all university work, so the warts are unlikely to appear. Instead of giving students a sense of the limits of Perl, they may have deputized another programmer who thinks Perl is the perfect hammer and all programming problems are nails.

Wirth, inventor of Pascal, concluded his 2008 "A Brief History of Software Engineering" retrospective with the statement that academia

has remained inactive and complacent. Not only has research in languages and design methodology lost its glamour and attractivity, but worse, the tools common in industry have quietly been adopted without debate and criticism. Current languages may be inevitable in industry, but for teaching, for an orderly, structured, systematic, well-founded introduction they are entirely mistaken and obsolete.

This is notably in accord with the trends of the 21st century: We teach, learn, and perform only what is immediately profitable, what is requested by students. In plain words: We focus on what sells. Universities have traditionally been exempt from this commercial run. They were places where people were expected to ponder about what matters in the long run. They were spiritual and intellectual leaders, showing the path into the future. In our field of computing, I am afraid, they have simply become docile followers. They appear to have succumbed to the trendy yearning for continual innovation, and to have lost sight of the need for careful craftsmanship.

If we can learn anything from the past, it is that computer science is in essence a methodological subject. It is supposed to develop (teachable) knowledge and techniques that are generally beneficial in a wide variety of applications. This does not mean that computer science should drift into all these diverse applications and lose its identity. Software engineering would be the primary beneficiary of a professional education in disciplined programming. Among its tools languages figure in the forefront. A language with appropriate constructs and structure, resting on clean abstractions, is instrumental in building artefacts, and mandatory in education. Homemade, artificial complexity has no place in them. And finally: It must be a pleasure to work with them, because they enable us to create artefacts that we can show and be proud of.[49]

The trend I discussed earlier, of language design moving from academia to corporate research labs, has continued, with language design then moving from corporate research labs to corporate business divisions, becoming an explicit part of a company's strategy by defining a platform for other programmers. Sun came out with Java (although Java has now migrated to be owned by a combined industry/user committee), Microsoft invented C#, and Apple, after some acquisitions, wound up owning Objective-C, which it more recently evolved into a language called Swift. There are advantages to this: when languages were designed by professors who had no particular interest in productizing the related tools, they often had many different flavors with different features and syntax, such as the many versions of BASIC that the *BASIC Computer Games* book had to accommodate. Having a single company providing the reference version of the compiler and driving the language forward does avoid that problem. But it also moves academia further away from industry in this critical area.

Given that academics no longer tend to be involved in language design, perhaps it is not surprising that they are less interested in comparing languages than they once were. Certainly universities have a limited amount of time to interact with students and a lot to teach them; it does not make sense for schools to be too hasty. The latest crop of languages should be given time to prove themselves before they are taught in school. Still, it is disappointing, as Wirth says, that schools are followers in this important area.

The industry, meanwhile, has done a lot of thinking about languages in recent years. But it has done even more thinking about something else: how to manage software projects, the topic of the next chapter.

9 Agile

Millions of years ago, powerful geologic forces in northern Utah converged to uplift the Wasatch Range and create a 160-mile north-south mountain chain that provides a dramatic backdrop to what is now Salt Lake City.[1]

In February 2001, another convergence occurred in the area, when seventeen men met at the Snowbird Resort to discuss a common enemy.[2]

Who were these men? They were an assortment of Americans, Canadians, and Britons, plus one Dutchman. The group included Dave Thomas, coauthor of the influential 1999 book *The Pragmatic Programmer*; Alistair Cockburn, author of *Writing Effective Use Cases*; Martin Fowler, author of *UML Distilled*; Kent Beck, inventor of extreme programming (and unit testing); and Ward Cunningham, who invented the wiki. These are all various software techniques that could be characterized as lightweight (except for the wiki, which is just cool).

Their enemy was software development methodologies with names like Rational Unified Process and Capability Maturity Model, which were gaining traction, for lack of any opposition, as potential best practices for software development (or at least, as the best practices for managing software development). They could be termed heavyweight—lots of writing specs up front and defining specific milestones to be checked off during the process. Despite their lofty names, they weren't making software development any more predictable.

Microsoft, certainly, had been beset by extremely public delays in the previous decade. The first version of Windows NT, on which I labored after joining the company in 1990, was initially estimated to take two and a half years but wound up taking nearly five.[3] Windows 95, which shipped to customers in August 1995, was originally planned to be Windows 93.

All seventeen men at the meeting in Utah had what they considered to be a better idea, although they didn't all have the *same* better idea. There had been cross-pollination, and many of them had worked or written together at various times, but they weren't all pushing the same idea so much as pushing against the same set of them. Some of the seventeen were talking about better ways to specify software, some about better ways to produce time estimates for it, some about better ways to design it, some about better ways to actually write it, and others about better ways to coordinate work. But they recognized that as seventeen separate outposts, they were not making much progress in the battle against heavyweight processes, and they decided that combining forces would give them more leverage.

During that 2001 meeting in Utah, the group adopted the word *agile* to unify their efforts. The term originated with Fowler and has a nice ring to it. Agile sounds better than whatever slow, stodgy term it is the opposite of, and certainly a lot better than *lightweight*; in the movie *The Karate Kid*, Ralph Macchio starts out lightweight, but ends up agile and overcomes his bullies.

The main output of the Snowbird gathering was the "Manifesto for Agile Software Development":

We are uncovering better ways of developing software by doing it and helping others do it. Through this work we have come to value:

Individuals and interactions over processes and tools
Working software over comprehensive documentation
Customer collaboration over contract negotiation
Responding to change over following a plan

That is, while there is value in the items on the right, we value the items on the left more.[4]

Agile is more of a branding exercise than any single approach, so a software development team announcing "we are Agile" doesn't mean much; it primarily signifies being au courant with progressive software development. Agile has oozed out into the world beyond software. My brother and sister, who work in old, well-respected, and time-proven fields (transportation engineering and scientific publishing, respectively), have over the past decade been hearing about Agile methodology and how it could help them.

The most well-known technique under the Agile umbrella is known as Scrum. The term comes from the scrum in rugby, where two teams link arms at the beginning of the match and run into each other in an attempt to get the ball (visualize "link arms," not "run into each other"). Scrum had been kicking around for about a decade before the Agile Manifesto was written, after being originally presented in a paper by Ken Schwaber and Jeff Sutherland at OOPSLA 1995.[5]

At its heart, Scrum is the Agile Manifesto mapped onto software project management. Programmers working on a Scrum team meet briefly every day to provide status and ask for help if needed, which precludes the need for any formal system to track dependencies between their work ("individuals and interactions over processes and tools"); aim to deliver new features in small increments rather than pieces of code that will only work once they are all completed and stitched together ("working software over comprehensive documentation"); rely on customer feedback on the delivered increments to figure out what to do next as opposed to planning a larger deliverable up front ("customer collaboration over contract negotiation"); and view changing customer requirements as a positive sign that people are using their stuff, rather than an excuse to complain ("responding to change over following a plan").

Scrum is about how to manage software projects, not about how to write the code. This is not a secret; the "Scrum Guide" website states, "Scrum is a process framework that has been used to manage complex product development since the early 1990s. Scrum is not a process or a technique for building products; rather, it is a framework within which you can employ various processes and techniques."[6] In fact, one key assertion behind Scrum is that there exists no solid process or technique to develop software, but that's OK; as Fowler writes, "A process can be controlled even if it can't be defined."[7] This is certainly in contrast to what Mills wrote a generation earlier: "My approach to software has been that of a study in management, dealing with a very difficult and creative process. The first step in such an approach is to discover what is teachable, in order to be able to manage it. If it cannot be taught, it cannot be managed as an organized, coordinated activity."[8]

It is often stated that Scrum replaced a software development process known as *waterfall*. Brooks, in his famous essay "The Mythical Man-Month," gives the following advice on splitting up time within a project: "⅓ planning, ⅙ coding, ¼ component test and early system test, ¼ system test [once] all components [are] in hand." His goal was to prod managers to allow more time for testing (and to a lesser extent, planning): "In examining conventionally-scheduled projects, I have found that few allowed one-half of the projected schedule for testing, but that most did indeed spend half of the actual schedule for that purpose. Many of these were on schedule until and except in system testing."[9] In other words, testing will extract its pound of flesh, whether you budget adequate schedule time or not; if you don't, then the project will slip.

What was implicit in that guidance was the one-way flow of the development process: first you plan, then you code, then you test each component, and then you test the whole thing together. This is where the word *waterfall* comes from, since the process is like water going over a fall. You don't reopen the planning process after coding has started, nor do you begin coding before the planning is complete. And you don't start testing until you reach "code complete" (fixing bugs found in testing can involve writing code, but the idea is to confine yourself to fixing bugs as opposed to making enhancements; in Knuth's framing of the distinction, you should always feel guilty, never virtuous). Brooks was arguing for changing the division of time between the phases, but not for changing the unidirectional transit through them.

In 1995, Brooks came out with a twenty-year anniversary edition of his book *The Mythical Man-Month*, in which the original essay of the same name had appeared. He included a new essay titled "*The Mythical Man-Month* after 20 Years," which states that "the waterfall model is wrong."[10] Brooks was responding directly to his own original essay "Plan to Throw One Away," which claimed that the first version of any system is going to be terrible, "too slow, too big, awkward to use, or all three," and "the only question is whether to plan in advance to build a throwaway, or to promise to deliver the throwaway to customers" (as he puts it, "Seen this way, the answer is much clearer").[11] Brooks had originally advocated building a "pilot system," that first terrible system, but never delivering it to your customer and instead starting over with the knowledge you have gained.

I never particularly liked this advice; it is un-engineer-y to not be able to build software that is usable the first time. Brooks did state that chemical engineering plants are built this way, with a smaller plant used to test a process, but presumably this is only done once for a given chemical process, not for every plant that uses the same process. Long bridge designs are based on information gathered from building shorter bridges, yet not every long bridge needs a specific shorter version constructed to prove that the real one won't fall down.

So I was glad to see that Brooks, with twenty years of hindsight, had changed his mind about "Plan to Throw One Away" (this was retroactive gladness years later; like most programmers, I wasn't reading his books in the 1990s, a period when Microsoft was heedlessly using a waterfall-ish approach). His main complaint was that his earlier advice accepted the waterfall model as fact: "Chapter 11 [the 'Plan to Throw One Away' essay] is not the only one tainted by the sequential waterfall model; it runs through the book, beginning with the scheduling rule in Chapter 2 [the 'Mythical Man-Month' essay]."[12]

"The basic fallacy of the waterfall model is that it assumes a project goes through the process *once*," Brooks continues,

that the architecture is excellent and easy to use, the implementation design is sound, and the realization is fixable as testing proceeds. Another way of saying it is that the waterfall model assumes the mistakes will all be in the realization, and thus that their repair can be smoothly interspersed with component and system testing.

"Plan to throw one away" does indeed attack this fallacy head on. It is not the di-agnosis that is wrong; it is the remedy. ... The waterfall model puts system test[ing], and therefore by implication *user* testing, at the end of the construction process. Thus one can find impossible awkwardness for users, or unacceptable performance, or dangerous susceptibility to user error or malice, only after investing in full con-struction.[13]

Rather than mimicking the smooth flow of water over a waterfall, the process was beginning to look more like riding a barrel over one: moments of sheer terror followed by getting crushed in the churn at the end.

Brooks mentions that the waterfall model was enshrined in a US Depart-ment of Defense specification for all military software. Back in the realm of users who pay for their own software, he recommends, "implementers may well undertake to build a vertical slice of a product, in which a very limited function set is constructed in full, so as to let early sunlight into places where performance snakes may lurk."[14] It's not just performance snakes; it's any issues the users might encounter. If you know your software will need a database layer, rather than start by implementing all the functionality you think you will eventually require, you write the minimal database support needed for a single user-visible feature, along with the minimal support needed in any other layers, and present that to the customer; that is the ultimate way you will determine if your database is designed properly. Then you go back and work on another feature: lather, rinse, repeat.

It's not just the users who prefer vertical slices; it's also the programmers. Brooks remarks that at one point during his tenure at the University of North Carolina, "I switched to teaching incremental development. I was stunned by the electrifying effect on team morale of that first picture on the screen, that first running system."[15]

Scrum focuses aggressively on delivering new functionality to the user as often as possible. The timeline for delivery is known as a *sprint*, and typically lasts two to four weeks. Around the sprint, Scrum has certain artifacts, including the *product backlog* (a list of all work that could be considered for future sprints) and *burndown chart* (which tracks the total estimated work remaining in this sprint, hopefully hitting zero at the end of the sprint—with the caveat that only items that are entirely complete can have their hours crossed off). At the beginning of each sprint, the team selects what it feels are the right product backlog items (right in terms of both "what the customer wants next" and "what we think will fit in one sprint") and then spends the rest of the sprint working to deliver those items. The term *sprint* is frequently misinterpreted to imply frenetic activity that leaves you burned out, but the goal is that teams can work sprint after sprint without breaks. A sprint should feel more like one mile of a marathon, not a mad dash to the finish line.

Since Brooks was writing about replacing the waterfall model with the incremental delivery of vertical slices in the same year that Scrum was being presented at OOPSLA, it might appear that Schwaber's delineation of Scrum represented the recording of an already-emerged consensus about how to move away from the waterfall model of development. Actually, Scrum was a much more aggressive departure.

Schwaber's paper was presented as part of a larger OOPSLA workshop titled "Business Object Design and Implementation," where a "business object" is a reusable software component that can be knitted together with other business objects to create applications—the familiar object-oriented dream. The Scrum paper doesn't have anything particular to do with this. The summary paper from the entire business object workshop merely states, "New systems will require that loosely coupled, reusable, plug compatible components be constructed using a tightly coupled development method that combines business process reengineering, analysis, design, implementation, and reusable component market delivery systems similar to today's custom IC chip industry." Yet it also includes the following summary of the Scrum paper: "The stated, accepted philosophy for systems development is that systems development process is a well understood approach that can be planned, estimated, and successfully completed. This is an incorrect basis."[16] Essentially, Schwaber was saying that the premise of the rest of the workshop was faulty; there was no "tightly coupled development method," nor will there ever be one.

The summary continues,

SCRUM states that the systems development process is an unpredictable, complicated process that can only be roughly described as an overall progression. SCRUM defines the systems development process as a loose set of activities that combines known, workable tools and techniques with the best that a development team can devise to build systems. Since these activities are loose, controls to manage the process and inherent risk are used.[17]

The mid-1990s, when Brooks wrote his updated essay and Scrum was getting started, was also the time that the book *Microsoft Secrets* came out. The book lays out various principles, as reported by Microsoft employees, for how the company handles development, including "work in parallel teams, but 'synch up' and debug daily," "always have a product you can theoretically ship," and "continuously test the product as you build it."[18] This sounds marvelously Agile, but having worked on a large Microsoft product in the early 1990s, I know that the techniques used were a far cry from what Scrum was advocating. Brooks, in his 1995 update, is impressed at learning that Microsoft builds and tests its software every night.[19] The reality is that while we did ensure that the software built every night—meaning that it produced a compiled program without hitting any errors—and did do minimal testing each day, it was a product we could "theoretically ship" only in the most theoretical meaning of the word *theoretical*. Our software was developed in *milestones* lasting six to nine months each, and although we did not follow a strict waterfall process during each milestone, we definitely back-loaded the testing, and only toward the end of any given milestone was the software reliable enough to release externally. Effectively our "sprint" duration, and the timeframe for the "slices" we delivered, was six to nine months—much longer than the two to four weeks that Scrum advocates. Even equating the milestones with "long sprints" is wrong, because Scrum explicitly contraindicates the "mini waterfall" approach to a sprint, where you might split your four weeks into a week of planning, two weeks of coding, and one week of testing. Every feature delivered during a sprint is supposed to be ready to ship to a customer on the day it is completed.

So while many companies had moved away from the pure waterfall model to something a bit more iterative (or never used pure waterfall to begin with), Scrum made for a dramatic acceleration of that movement.

Even more dramatic is the second-best-known Agile methodology, Extreme Programming (known as XP). Invented by Beck (who had earlier originated unit testing in its current form), XP is based on a set of rules around the planning, managing, designing, coding, and testing of software.[20] Planning and managing are Scrum-like, with daily meetings and a focus on frequent small releases, although XP is more prescriptive in some cases. The guidance on the design/coding/testing phases gets into actual specifics of how the software should be engineered, which Scrum ignores. The key to the approach is writing unit tests, and ensuring that those unit tests are run often, and writing new unit tests whenever new code is added or a bug is found that snuck past the current set of unit tests.

XP also mandates the somewhat-controversial practice of pair programming, in which two programmers work together at all times, sharing one computer; typically one is coding while the other is watching. The idea here is, quite literally, that two heads are better than one; the second programmer provides a continuous code review of what the first programmer is doing (having a second programmer watching also discourages unnecessary web-based diversions while you are supposed to be working, which no doubt has an effect on productivity).

XP does attempt to avoid some of the "religious" arguments about coding. It mandates that a coding convention be written down and adhered to—without specifying a particular coding convention, but at least ensuring that arguments over proper coding style will happen once, be resolved in some way, and thereafter not brought up again (tabs or spaces, just pick one and go with it). On the question of making your code flexible to anticipate future changes, XP is clear: don't do it. Write the code for the requirements of the feature you are working on now, and if the requirements change, because of a new feature or user feedback, modify the code then. Since the code will have good unit tests, you can make these future modifications without worrying about accidentally breaking something because your understanding of the code is not fresh in your mind. And until you have new requirements, you won't know what changes are needed, so it is foolish to attempt to anticipate them now. Any opinions to the contrary can be dismissed with the mantra "You Ain't Gonna Need It" (YAGNI).

Scrum and XP are designed for small teams; the daily meeting doesn't scale well beyond ten to fifteen people because the chance of any one programmer caring about another programmer's status decreases as the number of attendees rises. Schwaber and Sutherland's original 1995 OOPSLA paper starts out with an axiom: "Small teams of competent individuals, working within a constrained space that they own and control, will significantly outperform larger development groups."[21] It is hard to argue with this if performance is based on a metric like "code delivered per person"; it is generally understood, in areas far beyond software engineering, that the communication and coordination overhead will increase as you work on larger projects. Nonetheless, it is misleading to imply that small teams will produce more software than larger ones of any size.

Beck writes, "Size clearly matters. You probably couldn't run an XP project with a hundred programmers. Not fifty. Not twenty, probably. Ten is definitely doable," and later says, "If you have programmers on two floors, forget it. If you have programmers widely separated on one floor, forget it."[22]

Another problem is that when you are working on the first version of a piece of software, it can be hard to produce anything that users can use in two to four weeks, or even in small multiples of two to four weeks. Before Windows NT could start shipping public releases at the end of six- to nine-month milestones, it took several years to create anything that was usable at all, because so much of the internals of an operating system need to be written before it can handle a single request from a user.

Microsoft used to be *feature driven* in its products, meaning that teams would establish a planned set of features and then work until they were available, accepting whatever schedule slip was needed; at a certain point the company switched to being *date driven*, where teams would set a date and only include the features that could be completed by that date, cutting features in the middle of a project if they looked to be in danger. This made things much more predictable for customers. But this is a luxury that is available when you have an existing product; for the first version of Windows NT, the critical feature "the operating system works" could not be cut. Date-driven scheduling is heralded as a breakthrough in project management, but it's no coincidence that the switch away from being feature driven happened around the time that all of Microsoft's major products (Windows, Office, its compilers, the SQL Server database program, and the Exchange e-mail server) had established, working versions, which could then have individual features added on (or not) in subsequent versions.

In fact, the original OOPSLA Scrum paper states, "Scrum is concerned with the management, enhancement and maintenance of an existing product, while taking advantage of new management techniques and the axioms listed above. Scrum is not concerned with new or reengineered systems development efforts." By the time Schwaber's first Scrum book came out in 2002, this distinction had been lost, and Scrum was presented as applicable to both new and ongoing projects.[23]

As an aside, the OOPSLA paper states another axiom: "Product development in an object-oriented environment requires a highly flexible, adaptive development process," and later says, "Object Oriented technology provides the basis for the Scrum methodology. Objects, or product features, offer a discrete and manageable environment. Procedural code, with its many and intertwined interfaces, is inappropriate for the Scrum methodology."[24] I'm not sure what being object oriented has to do with it; procedural programming also required a highly flexible, adaptive development process. If you believe the loudest object-oriented supporters, procedural programming would require even more flexible processes since it is missing the special sauce that object-oriented programming provides. The vaguely implied equating of "objects" and "product features" makes me think this was either a sop to the OOPSLA crowd or reflection of the heady early days of the object-oriented frenzy. For what it's worth, I have read several books on Scrum (including Schwaber's, which makes no mention of this), become a Certified Scrum Master, and taught Scrum to teams inside Microsoft for several years, and I never heard that Scrum was unsuited to procedural programming or observed problems with Scrum that were unique to teams using procedural languages.

Mills once described courses available to programmers as "new names for common sense," and while common sense is better than a lack of it, Scrum is still tackling the easiest problem in software: small teams working for a single customer on incremental improvements to an already-functioning piece of software.[25] This is not to say it is not useful; teams in those situations were taking archaic approaches, such as using a waterfall model to deliver the complete solution before getting any customer feedback, and Scrum can certainly get them on a better path.

Where waterfall attempts to plan out the details of a project carefully and predict a completion date, Scrum states that the team will work diligently on pieces of a project (the "best that a development team can devise" from Schwaber's original paper—meaning, "trust us and stop nagging"), and in the right order, always focusing on delivering working code to the user at the end of each sprint. The Agile approach to preventing long schedules that slip is to avoid long schedules. As Schwaber and coauthor Mike Beedle explain, "Several studies have found that about two-thirds of all projects substantially overrun their estimates," and they address the "risk of poor estimation and planning" this way: "Scrum manages this risk ... by always providing small estimates. ... Within the Sprint cycle, Scrum tolerates the fact that not all goals of the Sprint may be completed."[26] Beck talks about "schedule slips—the day for delivery comes, and you have to tell the customer that the software won't be ready for another six months," and explains, "XP calls for short release cycles, a few months at most, so the scope of any slip is limited."[27]

In other words, these methodologies don't change the fact that software engineers are bad at estimating; they just keep the estimates short, so that even a significant slip, in percentage terms, is not that bad in calendar terms. To be fair, they do emphasize frequent delivery of working code to customers, which allows customer-driven course correction as needed and encourages team members to complete higher-priority work first, which is a step in the right direction (absent this nudge, they would tend to tackle the most interesting technical problem first). And Agile proponents point out, accurately, that if a team stays together and works on similar kinds of work, it will become better at estimation—but that is not new information or unique to Agile.

Socrates is quoted in Plato's *Apology* as saying, "I am wiser than this man; it is likely that neither of us knows anything worthwhile; but he thinks he knows something when he does not; whereas when I do not know, neither do I think I know; so I am likely to be wiser than he is to this small extent, that I do not think I know what I do not know." In this sense Scrum, which says that software projects are inherently uncontrollable, is wiser than the waterfall methodology, which proposes to control them without knowing how. Although Scrum proponents may want to heed a follow-on insight from Socrates: "The good craftsmen seemed to have the same fault as the poets: each of them, because of his success at his craft, thought himself very wise in other most important pursuits, and this error of theirs overshadows the wisdom they had."[28]

Agile is not really the opposite of waterfall, partly because true waterfall wasn't used much at the time that Agile came along. If you are looking for an approach that is as far from Agile as possible, it is the Personal Software Process (PSP) and Team Software Process (TSP), developed at the Software Engineering Institute (SEI), a software think tank at Carnegie Mellon University, under the guidance of Watts Humphrey, a longtime manager of software teams at IBM.

The PSP approach is laid out in the preface to Humphrey's 1995 book *A Discipline for Software Engineering*:

Society is now far too dependent on software products for us to continue with the craft-like practices of the past. It needs engineers who consistently use effective disciplines. For this to happen, they must be taught these disciplines, and have an opportunity to practice and perfect them during their formal educations.

Today, when students start to program, they generally begin by learning a programming language. They practice on toy problems and develop the personal skills and techniques to deal with issues at this toy problem level. ... These programming-in-the-small skills, however, are inherently limited.[29]

The PSP solution is to take practices used for large-scale software development—and Humphrey, at IBM, was managing some of the largest-scale software development of his day—and scale them down to work for single-person programs. Having programmers use these techniques on small programs would prepare them for proper large-scale software development (which is addressed in phase two of Humphrey's plan, the TSP). In this approach, the PSP is the exact opposite of Agile, which takes techniques optimized for small teams and then implies that they will work for larger teams.

Recognizing that this sounds a lot like exhorting people to eat their vegetables, Humphrey throws down a little challenge: "The PSP is a self-improvement process. Mastering it requires research, study, and a lot of work. But the PSP is not for everyone. Recall that the PSP is designed to help you be a better software engineer. Some people are perfectly happy just getting by on their jobs. The PSP is for people who strive for personal achievement and relish meeting a demanding challenge." He also includes a chapter on how to stay the PSP course even if you are the only person on the team using it, when your manager and coworkers are giving you funny looks.[30]

The PSP relies heavily on counting three things—lines of code, defects, and time—and performing various mathematical calculations among them, so as to enable predictions about the future. It makes use of formal code inspections—a group activity first proposed by Michael Fagan from IBM in 1976.[31] Formal inspections differ from the individual code reviews we encountered in chapter 3 in a variety of important ways: the inspectors are given time ahead of the meeting to read the code; guidelines on what to look for in reviews are created and kept updated; the meeting has a formal leader to keep things moving; and the results of the inspection (the number of defects found per line of code) are tracked and analyzed.[32]

My first experience with any kind of formal code review at Microsoft was back in 1993, when I was working on low-level networking code in Windows NT. A group of us sat down with printouts of my code and started walking through them. The date was January 20, the day of Bill Clinton's presidential inauguration, and the day a windstorm swept through the Seattle area and knocked out power to Microsoft. Undaunted, we gathered in a conference room near the window—and completely missed the fact that a tree had fallen on a catering truck outside our building, with the truck's cargo of pizzas subsequently being distributed free to everybody in the building, except for those of us hunkered down doing a code review in the fading light.

Beyond that trauma, I can see in retrospect that this was nothing like an inspection is supposed to be. Nobody had read the code ahead of time, and the results were not tracked; it was more like a set of parallel individual code reviews based on whatever surface-level faults could be spotted. In a report reprinted in a book by Tom Gilb and Dorothy Graham, these individual, ad hoc code reviews were described as "the least effective, but most used, of all defect removal techniques."[33] Meanwhile, SEI (and Gilb and Graham, for that matter) has research demonstrating real improvements from code inspections.

On the other hand, like Agile, the PSP doesn't say much about how to write code—how to put line B after line A. The closest it gets to talking about actual software design is to mention that both top-down and bottom-up designs can be useful in different situations, as can starting in the middle, and that focusing on vertical slices can be a good idea, but so can building the entire system up in layers.[34] Since it's the PSP, you spend time thinking about your strategy up front and also ruminating about whether it went well afterward, which is better than blindly using the same strategy as the last time, but doesn't provide much guidance on how to proceed in a new problem area. Scrum at least gives you concrete guidance, to concentrate on vertical slices, which may be bad advice in a given situation, but at least prevents dithering—since in the end, given the one-off nature of most projects, it is hard to know if a different design strategy would have worked better.

I was never exposed to the PSP until I worked in Engineering Excellence at Microsoft and we taught some of its concepts in our courses, but I can see why programmers would instinctively recoil from it. Thinking about all that tracking, just for my own personal improvement, makes my head ache. Thinking about the PSP book makes my arm muscles ache; at over 750 pages, it's the longest software engineering book I know (the TSP book clocks in at a relatively svelte 450 pages).[35] It spends an entire 30-page chapter talking about how to count lines of code (admittedly a subject of debate in programming circles). Per PSP, you are supposed to count every syntax error that the compiler catches as a defect, which is duly logged for future analysis. Fixing compilation errors is a mechanical, annoying task, but one that doesn't take that long; adding in the mechanical, annoying task of logging the errors makes me shudder (and you are further supposed to classify the errors into one of about twenty different categories).[36]

There is even debate in PSP circles about whether it makes sense to do a code review *before* you compile the code the first time, which instinctively seems like a waste of time. Why spend time looking for errors that the compiler can catch in a few seconds? Ah, says the PSP data, but around 10 percent of errors that you would expect the compiler to catch are not caught because they inadvertently form valid syntax, and those are particularly sneaky bugs to figure out later.[37] And having all the bugs available to be found in a pre-compiler review makes the code a more target-rich area, which makes the code review more rewarding and hence more likely to be taken seriously.

It all does make a certain sense; classifying compiler errors by type could be a lot of work, but if I realize that I make specific kinds of mistakes more than others, I can focus on avoiding those and become more efficient, and yet. ... I don't consider myself somebody who is "just happy getting by on their jobs," in Humphrey's accusatory phrase, but I can well understand why in Engineering Excellence our Scrum courses had much better uptake than our PSP-inspired estimation and inspection courses, both of which I felt were more relevant inside Microsoft than Scrum training. When you've achieved a level of success being self-taught, it is much easier to accept a methodology like Scrum, which says that even the limited amount of tracking you are being asked to do is unnecessarily hobbling you, than one like PSP, which says that you need to do more of it.

For PSP to take hold as the natural order of things it would be helpful for it to be instilled early on, but for many programmers, "early on" is during high school; few people trying to hack out a quick mobile app or website are going to worry about something like PSP, if they have even heard of it. Also, the idea built into the PSP of taking processes appropriate to large-scale projects and scaling them down to small programs, where they are not needed except as training for future work on large programs, makes PSP a tough sell. It is not even taught to undergraduates at Carnegie Mellon, the university with which SEI is associated.

Fundamentally, Scrum in particular and Agile in general are optimistic: assume things will go well, trust your team, and fix the process if needed. PSP and other command-and-control techniques are pessimistic: assume things will go terribly wrong unless you invest a significant percentage of your time in preventing problems. As a manager, I much preferred taking an optimistic approach, and I think the people who worked for me appreciated it too. But we still did a lot of planning and tracking that went well beyond what any Agile methodology would recommend.

Agile is correct in recognizing that trying to figure out up front how long a software project will take, given our current techniques for both estimation and software engineering, is a fool's errand. A favored tactic of managers, before Scrum and XP gave programmers the cover needed to tell them to butt out, was to ask a programmer for an estimate in the early days of a project, when they didn't yet know enough about its details to be able to make a decent estimate, and then hold the programmer to that seat-of-the-pants estimate. In a book on software estimation (appropriately subtitled *Demystifying the Black Art*), Steve McConnell writes about the *Cone of Uncertainty*: the fact that the error range for software estimates starts out large, yet gradually shrinks as you begin to do more investigation into how you will implement the work, and shrinks even further after coding begins.[38] Asking a programmer to provide an estimate at the widest part of the cone, and then never revisiting it, is the worst-possible approach. (He further points out that having individual programmers supply estimates, rather than trying a "wisdom of the crowds" approach of asking multiple people for an estimate even if they won't be the ones doing the work, further contributes to the inaccuracy of estimations, no matter at what point on the cone they are done.)[39]

In *The Soul of a New Machine*, his Pulitzer-Prize-winning book about the engineering of a new minicomputer at a company called Data General in the late 1970s, Tracy Kidder describes how early estimates became etched in stone:

There was, it appeared, a mysterious rite of initiation through which, in one way or another, almost every member of the team passed. The term that the old hands used for this rite—West invented the term, not the practice—was "signing up." By signing up for the project you agreed to do whatever was necessary for success. You agreed to forsake, if necessary, family, hobbies, and friends—if you had any of these left (and you might not if you had signed up too many times before). From a manager's point of view, the practical virtues of the ritual were manifold. Labor was no longer coerced. Labor volunteered.[40]

In a 2000 study of a high-tech company, Ofer Sharone referred to this situation, in which employees self-impose the type of workplace pressure that one would expect to come from their managers, as one part of what he calls "competitive self-management," which can "engender intense anxiety among [the company's] engineers regarding their professional competence."[41] The other part is grading employee performance on a rigid curve; I don't know if this was in effect at Data General, but it certainly is at a lot of software companies.

The Soul of a New Machine was about hardware development, but it still provides the most accurate depiction I have ever read of what it is like to work on a large "version 1" software project. Creating the first version of a new computer (it was Data General's first 32-bit minicomputer) is an all-or-nothing endeavor; you can't ship half a computer. The result was high pressure and long hours, including this portrait of "signing up" in action, between a programmer named Dave Epstein and his boss, Ed Rasala:

Some weeks ago, Ed Rasala asked Epstein, "How long will it take you?"
    Epstein replied, "About two months."

"Two months?" Rasala said. "Oh, come on."

So Epstein told him, "Okay, six weeks."

Epstein felt as if he were writing his own death warrant. Six weeks didn't look like enough time, so he's been staying here half the night working on the thing, and it's going faster than he thought it would. This has made him so happy that just a moment ago he went down the hall and told Rasala, "Hey, Ed, I think I'm gonna do it in four weeks."

"Oh, good," Rasala said.

Now, back in his cubicle, Epstein has just realized, "I just signed up to do it in four weeks."

Better hurry, Dave.[42]

There is a joke about software: "The first 90 percent of the work takes 90 percent of the time. The last 10 percent takes the other 90 percent." Since estimates almost always grow rather than shrink, as new unaccounted-for work is discovered during the implementation, asking programmers for an estimate early on, when the Cone of Uncertainty is at its widest, winds up committing the programmer to a schedule that is too aggressive. Yet managers can justify it because they are doing the noble thing and building up a schedule from the programmers' own estimates as opposed to mandating one from above.

Schwaber and Beck had presumably felt the pain of this. Schwaber and Beedle are clear that estimates are only for planning out a sprint and are not considered binding.[43] Beck mandates a forty-hour week, with few exceptions: "The XP rule is simple—you can't work a second week of overtime. For one week, fine, crank and put in some extra hours. If you come in Monday and say, 'To meet our goal, we'll have to work late again,' then you already have a problem that can't be solved by working more hours."[44] (Epstein, in the excerpt from *The Soul of a New Machine* above, did finish his project in the four weeks he signed up for, although clearly working more than forty hours a week.)

I confess that when I read *The Soul of a New Machine*, rather than being turned off by the descriptions of signed-up engineers working crazy hours, I wanted to be involved in such a project. It wasn't just the notion of going out in a blaze of glory. A project like that, given the urgency to deliver, promised a freedom to do what was needed, bypassing whatever rules or conventions you felt were in the way. It was the same freedom that Agile promises to programmers—it's just that we spent a lot more time savoring that freedom than Schwaber, Beedle, and Beck recommend. I was eventually on such a project, working on the first two versions of Windows NT from 1990 to 1994. But once that was completed, I was done with signing up and looked for mellower projects within Microsoft. Despite the impression from a story like *The Soul of a New Machine* that working crazy hours is a heroic undertaking, when it happens with software, it produces code that is rushed and poor quality, with a long tail of bugs for customers to uncover. In particular, the temptation is great to gloss over the handling of error cases: code that rarely runs, but is the most critical part when it is needed.

The Agile approach of only providing short estimates and not holding programmers' feet to the fire is clearly better than offering long estimates, stamping them on programmers' foreheads, and then missing the overall schedule anyway. Yet if you step back, this glosses over a more fundamental problem. Scrum is not a progressive way of managing software projects; it's a logical reaction to the current state of software development, which attempts to contain the damage by not overpromising to customers. Some version of waterfall is the way engineering projects should work; it's what any "real" engineering project is aiming to achieve, because ideally you would know enough to anticipate issues and plan accordingly, and be able to schedule out the work accurately based on previous experience on similar projects. Valuing "responding to change over following a plan" is another way of saying "don't expect me to be able to predict what I will get done," which is the current reality, but I hope things don't stay that way. Because while some change is due to customers seeing the software and realizing they don't like it, a lot is due to realizing that your internal implementation details, which the customer can't see, need to be reworked—and being unable to recognize ahead of time that you are going down the wrong path makes software development unpredictable.

You may have read about Scrum, the product backlog, and the burn-down chart, and thought, "Hey, I know nothing about software, but those things make sense." Which emphasizes the fact that Scrum has nothing to say about how to actually engineer software; it is focused on getting customer feedback through rapid iterations. Mary Shaw wrote about quote-unquote software engineering back in 1990 that "unfortunately, the term is now most often used to refer to life-cycle models, routine methodologies, cost-estimation techniques, documentation frameworks, configuration-management tools, quality-assurance techniques, and other techniques for standardizing production activities. These technologies are characteristic of the commercial stage of evolution—'software management' would be a much more appropriate term."[45] Scrum fits right into the software management category, not the software engineering category.

I do give Agile credit for acknowledging one important fact about programming, which previous methodologies tended to ignore: code is read a lot.

The significance of code reading has not been completely missed in the literature. The IBMers' *Structured Programming* book has a long chapter on code reading, complete with case studies, which begins, "The ability to read programs methodically and accurately is a crucial skill in programming. Program reading is the basis for modifying and validating programs written by others, for selecting and adapting program designs from the literature, and for verifying the correctness of one's own programs."[46] Mills, in *Software Productivity*, has a chapter titled "Reading Code as a Management Activity" (from 1972, thus predating his coauthorship of the *Structured Programming* book). He anticipates "a new possibility in PL/I: that programmers can and should read programs written by others, not in traumatic emergencies, but as a matter of normal procedure in the programming process."[47] Weinberg, in *The Psychology of Computer Programming*, also has a chapter on reading programs.[48] I'll mention in passing that my previous book, *Find the Bug*, talks about how to read code too.[49] In *Microsoft Secrets*, an engineer on Excel praises Hungarian for its salutatory effect on code reading: "Hungarian gives us the ability to just go in and read code. ... Being fluent in Hungarian is almost like being a Greek scholar or something. You pick up something and you can read it."[50] The actual content in that quote is somewhat divergent from reality, but it does show that reading code was an activity that programmers did and cared about, and tried (in vain, in this case) to make easier.

In classic waterfall programming, the goal was to write the code for the entire system you were going to produce and then hand it off to testing. If bugs were found, the programmer might experience the tribulation of revisiting the code, but it would be considered normal and even a positive sign if it were never looked at again.

In Agile, with its focus on delivering small increments of functionality under the YAGNI banner, it is understood that the code will be modified eventually, when "You Do Need It"; new vertical slices are written that touch existing code, and better ways to arrange the resulting whole are discovered in a process known as *refactoring*. Code is not meant to be written once and then never touched again; far from it. Rather than interpret this as "we got it right the first time," it would be seen as "we are probably not responding to our customers, and our code is getting moldy." As Beck put it, "A day without refactoring is like a day without sunshine."[51] The recognition that code is going to be read and modified often is an important mental switch from older methodologies.

The most extreme refactoring-based Agile approach is known as test-driven development, which mandates not only that unit tests be written for all code but also that unit tests be written first, before the code they are going to test. Furthermore, you strictly alternate between writing one test and writing the code to make that test pass—with no peeking ahead! YAGNI is the mantra, so if you were writing the code to score a game of bowling (which is the standard test-driven development example), and your first unit test involved a game that was all strikes, your actual product code at that point should look something like this:

```
int ScoreGame(Board b) {
    return 300;
}
```

Do you see what I did there? Naturally once you wrote a second unit test, which tested something other than a game of all strikes, you would need to write code to actually score the game based on the Board parameter, not just hard code a score of 300 that satisfied the first test.

My biggest concern about Agile is that it currently dominates the programming methodology discussion while only covering a narrow subset of the problems that software engineers can hit. In this entire chapter, that one-line method was the only code sample needed. Despite this, Agile is pitched as the savior of programming projects; as Schwaber and Beedle wrote in their book, "The case studies we provide in this book will show that Scrum doesn't provide marginal productivity gains like process improvements that yield 5–25% efficiencies. When we say Scrum provides higher productivity, we often mean several orders of magnitude higher, i.e., several 100 percents higher."[52] The actual case studies are underwhelming, to say the least (especially since they are handpicked, not controlled experiments), but Scrum still sells to an eager audience of programmers.

Earlier, I discussed the shift in the origin of software ideas, starting with universities in the early days, then moving to corporate research labs, and then turning to corporate product groups. Agile is the next evolution of this trend. Although its inventors began as programmers whose business was writing programs, they quickly morphed into consultants whose product was Agile knowledge itself. As with much other advice to programmers, Agile was not based on any research studies or empirical observation beyond what people noticed in their own work.

Academia, in particular, has almost nothing to do with Agile. It's easy to see why: with its new terminology and overhyped promises, Agile can come across as a fad, which universities would want to avoid. This in turn makes universities seem slow and stodgy to Agile practitioners, and possibly to new college graduates who fall under the sway of Agile. What is missing, to mend this gap, is more research on when exactly Agile practices are helpful, and when they are not. A methodology such as test-driven development is no doubt useful in some situations, but not in every one, which it perforce is proposed as a solution to. This provides ample ammunition for both sides of any argument.

In 2007, Scott Rosenberg published the book *Dreaming in Code*, in which he embedded himself with a well-credentialed group of programmers trying to write version 1 of an application: a personal information manager named Chandler. He unfortunately happened on a somewhat-dysfunctional team. Many of the members were tainted by previous success in that they were unable to distinguish factors that had legitimately contributed to that success from factors that didn't matter or were actively wrong. They all believed, in different ways, in the fallacy "if we just do this one thing, then the normal complications of software development won't apply."

Nonetheless, Rosenberg had enough personal experience with software to realize that their behaviors were not completely atypical. At a certain point he throws up his hands: "As I followed Chandler's fitful progress and watched the project's machinery sputter and cough, I kept circling back to the reactions I'd had to my own experiences with software time: *It can't always be like this*. Somebody must have figured this stuff out." He then spends a chapter wandering around some of the same back alleys I have covered here, including design patterns, XP, and PSP. Rosenberg concludes, "I can't say that my quest to find better ways of making software was very successful," but qualifies this by saying, "I don't think the methodology peddlers are snake oil salespeople."[53] It's just that the solutions being proposed don't help with a large, complicated project like Chandler. Rosenberg eventually got tired of waiting, and his book appeared before Chandler did.

One of the most recent flavors of Agile is the Software Engineering Methods and Theory (SEMAT) initiative, created "to identify a common ground for software engineering ... manifested as a kernel of essential elements that are universal to all software development efforts." SEMAT is introduced in a book subtitled *Applying the SEMAT Kernel*, but with a more ambitious title, *The Essence of Software Engineering*. Like any good jeremiad, it features a call to action, which states:

Software engineering is gravely hampered today by immature practices. Specific problems include:

– The prevalence of fads more typical of a fashion industry than of an engineering discipline
– The lack of a sound, widely accepted theoretical basis
– The huge number of methods and method variants, with differences little understood and artificially magnified
– The lack of credible experimental evaluation and validation
– The split between industry practice and academic research[54]

As I've written in similar situations, it's hard to argue with all that. How does SEMAT propose to address this? Not, initially anyway, by actually doing any experimental evaluation and validation. As programmer and former professor Greg Wilson comments in his "Two Solitudes" keynote talk from the SPLASH 2013 conference, the SEMAT book doesn't cite a single empirical study.[55] Instead, between the three forewords and twelve pages of testimonials at the end, it attempts to abstract out the common parts of software process management methodologies into a metamethodology, which could then be used to diagnose flaws in your actual methodology. Given that Agile is already somewhat removed from the actual problems of software engineering, taking a step back does not get you any closer to the essence.

Yet it does bring up a point about Agile. To some people, the problems with software relate to process management: making sure the requirements are correct, stakeholders are involved, and right team is in place. The actual writing of the software is an exercise left to the reader. For this audience, something like SEMAT *is* moving closer to the essence of software engineering.

I am not minimizing the importance of all that. The Agile techniques originated with consultants who were working on contract work; it helps you concentrate on the customer when you won't get paid unless they like what you deliver. This customer focus had often been ignored by programmers, who viewed customer change requests as evidence of their fickle "luserness," not a necessary step in making them happy. In Engineering Excellence at Microsoft, we studied a field known as human performance improvement, to be used in analyzing Microsoft teams. One of the key tenets of human performance improvement contends, "Put a good performer in a bad system, and the system will win every time."[56] In other words, the inputs people get from the environment have a greater impact on their performance than what they themselves bring to the table. Having good requirements and involved stakeholders is a critical part of the environment in which programmers operate.

But for a lot of software engineers, this has already been decided; the spec is written, the team is chosen, and now code needs to be written. Agile tends to peter out just as the engineering gets complicated. If you have a team that can meet every day in a room, its project is small enough that it can test most of its work via unit tests; and if the team stays together for the duration of a project, it won't hit mysterious problems calling unclear API because the person who wrote the API is probably in the room with the team. Schwaber and Beedle, discussing the issue of other people needing to learn the code that the team has written, came up with a simple yet impractical solution: "[We] instituted the following policy: whoever writes code owns it forever."[57] I suppose from the team's perspective the code can be owned forever, if their perception of time ceases once they stop working on it, in a sort of reverse big bang.[58] Unfortunately, customers don't have this luxury.

The complications of software happen at larger and longer scale than those sorts of projects. While Agile may make easy problems a bit easier, it doesn't help with the hard problems. It's appealing to programmers, but to make software engineering more of an engineering discipline, something else is needed.

10   The Golden Age

If you're like me, you dream of a day when software engineering is studied in a thoughtful, methodical way, and the guidance given to programmers sits atop a foundation of experimental results rather than the shifting sands of individual experience. Perhaps with a time machine, it would be possible to travel into the future and live in such a world.

Somewhat surprisingly, there is another way. It still requires a time machine, but you would point it in the opposite direction, toward the past. About forty-five years in the past to be precise.

After alighting in the early 1970s and locating the nearest computer bookstore, you would discover that you were in the middle of a fertile time for software engineering research. Books from that era wrestle with every problem that confronts us today, despite the fact that this period predates almost every piece of software still running. The first work on UNIX began in 1969; C was invented in 1971. Essentially everything that came before— mainframe systems running programs written in languages like COBOL and Fortran—has been replaced, with the Y2K crisis providing the final nail in many coffins. Since the software from that era is functionally obsolete, it is tempting to dismiss research from the same time period as equally outdated.

This would be a mistake. Today we have faster hardware, more expressive programming languages, and better debugging tools. But if you read the old books, it is clear that the fundamental issues have not changed. People need to learn to program, they write a lot of code, it doesn't integrate with other code, debugging it is hard, new programmers don't understand it, and so on. The software wasn't as complicated as the largest programs now, but the languages and tools were also more primitive, so the difficulty was about the same—presumably at a roughly equivalent position on the spectrum of human cognitive demand.

What is different is that back then, there were a group of people in academia and industry that was taking a systematic approach to figuring out the problems as well as how to solve them. This was the period just after the NATO conferences, the paint was still drying on the term *software engineering*, and the discipline was being investigated the way other engineering disciplines had been investigated in the past.

Consider the 1971 book *Debugging Techniques in Large Systems.*[1] The title would attract interest today: we work on large software systems and have to debug them. The book is not a monograph by one person; it's a collection of papers pulled from a conference held in summer 1970 at the Courant Institute of Mathematical Sciences at New York University—the first in a planned annual series. The participants were from both the academic and industrial communities (IBM figures heavily in the industry representation), and the conference was supported by a grant from the mathematics program of the Office of Naval Research.

I own a reasonably complete collection of modern books on debugging, which I pulled together when I wrote *Find the Bug*, my contribution to the corpus. But none of those books is the result of a symposium; they are all in the "here are some things I figured out while working with code" vein that is so prevalent in modern software books. It's not that the topics have changed much in the intervening decades. *Debugging Techniques in Large Systems* covers compilers that can catch bugs, how to design software to reduce errors, better debugger tools, how software should be tested to reliably find bugs, and the ever-elusive issue of proving program correctness.

Ironically, a lot of the advice exchanged back then was much harder to take advantage of than it would be today, because everybody was using different computer systems that ran incompatible software. It wasn't a question of being able to come home from a conference with a new tool that you could use immediately; instead you would have an idea of how you could improve the tools available on your own system, if you chose to undertake the task. Nonetheless, there was great interest in sharing knowledge for the sake of advancing the software engineering discipline.

Unfortunately, the excitement didn't last. When I started college in 1984, *Debugging Techniques in Large Systems* book was only thirteen years old, but for whatever reason I never was exposed to it or any other book about software from this era except for the original *The C Programming Language* reference.

I had barely heard of the software researchers who were active back in the day. I knew the sound bite version: Brooks was known for saying that "adding people to a late software project makes it later," and Dijkstra was the guy who said "GOTO statement considered harmful." I was oblivious to how much of what they had written about software was still relevant and would continue to be. Those who ignore history, as they say, are doomed to repeat it. If you actually read Brooks's *The Mythical Man-Month*, from which I have quoted extensively in this book, he writes about documentation, communication, roles on teams, estimation difficulties, scalability of teams, code comments, and cost of code size—all things that the industry has struggled with ever since. The book came out in 1975, the year that Microsoft was founded—and yet we knew nothing of it!

Then there is Mills, one of the greats of the era whom I had never heard of until I started doing research for this book. Reading through *Software Productivity*, a collection of Mills's essays written between 1968 and 1981, you are treated to a preview of almost everything that has been debated about software since then: the different roles in software, how to design software, how to test it, how to debug it, unit testing, documentation, and so on. Mills was also a bomber pilot in World War II and created the first National Football League scheduling algorithm. To be fair, he was widely read at the time and had a successful career at IBM. After he died in 1996, the IEEE created the Harlan D. Mills Award for "long-standing, sustained, and impactful contributions to software engineering practice and research through the development and application of sound theory."[2] I can't recall ever hearing news about anybody winning it, however (for the record, Parnas and Meyer are both past honorees).

Many other researchers took a scientific approach to studying programming and programmers. Harold Sackman, in 1970, studied (among other things) the question of how much better good programmers are than bad ones.[3] Maurice Halstead, in 1977, examined whether the difficulty of a given programming problem could be quantified mathematically.[4] Mills wrote a paper with Victor Basili, one of the pioneers in the field, on techniques for documenting programs so programmers could understand them quickly.[5] The book *Studying the Novice Programmer* collects a series of studies on how people learn to program, covering such topics as how programmers understand the concepts of variables, loops, and IF statements, plus the personally relevant topic "A Summary of Misconceptions of High-School BASIC Programmers" ("The students' apparent attributing of human reasoning ability to the computer gave rise to a wide variety of misconceptions").[6]

The best part about all this research, besides the fact that it involved collaboration between industry and academia, is that the authors actually told you how to write software. No more endless checklists or theories on how to manage around the mess of software development; this is truly practical advice. Do this instead of that to make your code easier to debug; do that rather than this to make it easier for another programmer to read.

You want to settle the never-ending religious debates that continuously roil the waters of software development? Since all these have purported benefits, why not have two groups of people work with code written in different styles, and see how it affected the initial comprehension, ease of modification, and general maintainability of the code? Why not indeed! Here's Ben Shneiderman in *Software Psychology*, published in 1980, testing students on how different commenting styles affect readability of the same FORTRAN program.[7] Do mnemonic variables names matter? Larry Weissman did a series of experiments on that, reported in 1974.[8] What about indentation? Tom Love and Shneiderman were just a few of the people in the 1970s who investigated whether it helps readability.[9] Is GOTO really harmful? Max Sime, Thomas Green, and John Guest looked into that in 1973, as did Henry Lucas and Robert Kaplan in 1976.[10]

The battle over flowcharts is a rare example of how an engineering discipline is supposed to evolve. The flowchart approach has the programmer lay out every bit of control logic—every IF statement, loop, GOTO, and so on—on a diagram before coding it up. Flowcharts can work for basic decision trees. They are the ancestors of those diagrams on the last page of *Wired* magazine about "Should I Do X" or "What Sort of Y Should I Buy"— all those diamond shapes for decision points with an arrow leading away for each choice. Even the visual language, the shape of the boxes, is the same as software flowcharts. They were advocated in books such as Marilyn Bohl's 1971 *Flowcharting Techniques*, and continued to grace the pages of various how-to books for a while.[11]

The problem with flowcharts is that they don't help with the tricky parts of program comprehension. When reading an IF statement, the issue is not realizing that there is an IF statement in the code but instead knowing whether the IF logic is correct, which is as easy to read from the code as from a flowchart. Think back to our "Did the donkey hit the car" statement in DONKEY.BAS from chapter 1:[12]

```
1750 IF CX=DX AND Y+25>=CY THEN 2060
```

You could redraw this as a flowchart:

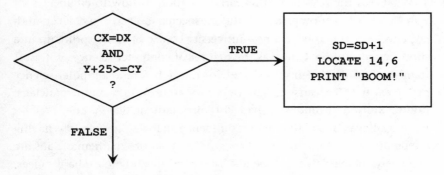

Figure 10.1
"Did the donkey hit the car?" in flowchart form

But that doesn't help us understand whether the IF test (CX=DX AND Y+25>=CY) is correct, which is where bugs might lurk. And for what it's worth, you don't know if a flowchart has been kept in sync as the code has been twiddled by intervening owners, so in the end you have to read the code anyway.

Flowcharts were eventually debunked, based on studies by Richard Mayer, the omnipresent Shneiderman, and others.[13] Luckily, this meshed with the empirical feedback from programmers that they were far more trouble than they were worth for anything of larger scope than a Microsoft interview question. Brooks, in a section of an essay titled "The Flow-Chart Curse"—with a title like that, do I need the quote? I'll provide it anyway as another illustration of how ideas that may work well for small programs break down when applied to large ones—observed that "the flow chart is a most thoroughly oversold piece of program documentation. ... They show decision structure rather elegantly when the flow chart is on one page, but the overview breaks down badly when one has multiple pages, sewn together with numbered exits and connectors."[14]

But that's about the only case I can think of where a once-fashionable programming methodology has been retired based on research—and even then I suspect it was primarily programmer lassitude, not the research studies, that led to flowcharts' extinction (I can personally recall both being advised to use flowcharts and later deciding on my own that they were a waste of time).

What was the reason for this early attention to how to engineer software? It's hard to know exactly, but I can speculate. There is a chicken-and-egg effect when you have a new university course of study spring up in a short period of time. How was the first set of computer science professors trained when they themselves went to college before a computer science major existed? The answer was that they were mostly mathematicians; Knuth, Mills, and Brooks all had PhDs in mathematics. As the son of a mathematician, I can state with confidence that despite possibly having a reputation for thinking deep thoughts in isolation, mathematicians are extremely collaborative and spend a lot of time meeting to exchange ideas, almost always building on the work of those who have gone before.[15]

Furthermore, when software first became a product that could be sold to customers, it was hardware companies that were writing the software; there were no "software-only" companies like Microsoft was in its early days. Each company was producing its own hardware that was incompatible with other hardware, and it needed an operating system to run the software along with compilers and tools to allow others to write software. Customers didn't buy computers to heat their office; they needed software for the specific problem they were trying to solve, and who better (or who at all, really) was there to write this than the company that also made the hardware? IBM, which is historically thought of as a hardware company, had to write a lot of software in order to sell its machines. SABRE, the original computer-based airline reservation system that American Airlines rolled out in the early 1960s, was written by IBM as part of a combined hardware/software deal with the airline. Mills worked in the Federal Systems Division at IBM, tasked with writing software customized for government customers.

When designing hardware, a company is doing "real" engineering: electrical engineering has built-up knowledge about circuit design, heat dissipation and power, and other topics that can't be solved with a "this worked for me last time" approach. Companies have to rely on research, both from academia and industry. In addition, you can't easily make changes late in the design of hardware the way you can with software; up-front design is worth the time. Presumably a hardware company would approach a software problem with the same disciplined approach.

Given that, it is understandable that early software engineering, driven by a combination of mathematicians in academia and hardware companies in industry, started down the path taken by other engineering disciplines, and you can see the results of this in the literature produced during that time. An observer in 1975 would have had reasonable confidence that the trend would continue, and that in a few decades, things would be figured out, codified, and then taught to students and reinforced through professional training.

That is not the way it turned out, to put it mildly. What happened?

In *The Psychology of Computer Programming*, Weinberg theorizes about a change caused by the arrival of terminals (this was back in 1971). When he talks about terminals, he means typing at a console that is still connected to a mainframe computer but allows you to edit and run programs interactively—the same rig I used to connect to McGill's computer from my parents' room circa 1981, except possibly with a screen display rather than a printer. This is a significant advance over older systems, where to run a program you had to submit it in person as a stack of punched cards, and then wait awhile for it to be scheduled and run, with your output being delivered to you by an operator who had access to the actual computer. Weinberg is discussing reading code as a way to improve yourself as a programmer and lamenting that this is done less than in the past:

With the advent of terminals, things are getting worse, for the programmer may not even see his *own* program in a form suitable for reading. In the old days—which in computing is not so long ago—we had less easy access to machines and couldn't afford to wait for learning from actual machine runs. Turnaround was often so bad that programmers would while away the time by reading each others' programs. ...

But, alas, times change. Just as television has turned the heads of the young from the old-fashioned joys of book reading, so have terminals and generally improved turnaround made the reading of programs the mark of a hopelessly old-fashioned programmer. Late at night, when the grizzled old-timer is curled up in bed with a sexy subroutine or a mystifying macro, the young blade is busily engaged in a dialogue with his terminal.[16]

Wading past the imagery in the last sentence, and ignoring his offhand use of a male personal pronoun as a stand-in for *programmer*, Weinberg's point is that programming with interactive terminals moves you away from the slower approach that characterized early software development, where you spent more time up front making it right, because the delay in running it was so much greater (and you had more time to chat with other programmers while standing around waiting for the operators to deliver your results). That was more like hardware engineering, where fixing a problem becomes so much more difficult once you have built physical hardware.

There was another seed germinating at this time that contributed to things veering away from the predicted path. The 1968 NATO conference in Garmisch, Germany, is remembered for the origin of the term *software engineering*, and the agreement between academia and industry that something needed to be done. Less remembered is the second NATO conference in Rome the following year, which did not end with the same feeling of togetherness. John Buxton and Brian Randell, editors of the proceedings, wrote the following:

The Garmisch conference was notable for the range of interests and experience represented among its participants. In fact the complete spectrum, from the inhabitants of ivory-towered academe to people who were right on the firing-line, being involved in the direction of really large-scale software projects, was well covered. The vast majority of these participants found commonality in a widespread belief as to the extent and seriousness of the problems facing the area of human endeavor which has, perhaps somewhat prematurely, been called "software engineering." …

The intent of the organizers of the Rome conference was that it should be devoted to a more detailed study of technical problems, rather than including also the managerial problems which figured so largely at Garmisch. However, once again, a deliberate and successful attempt was made to attract an equally wide range of participants. The resulting conference bore little resemblance to its predecessor. … A lack of communication between different sections of the participants became, in the editors' opinions at least, a dominant feature. Eventually the seriousness of this communication gap, and the realization that it was but a reflection of the situation in the real world, caused the gap itself to become a major topic of discussion. Just as the realization of the full magnitude of the software crisis was the main outcome of the meeting at Garmisch, it seems to the editors that the realization of the significance and extent of the communication gap is the most important outcome of the Rome conference.[17]

In other words, once people started to get away from broad recognition of the problem and into details of potential solutions, the gap between academia and industry began to manifest itself. Roger Needham and Joel Aron addressed this difference in a working paper at the second conference:

The software engineer wants to make something which works; where working includes satisfying commitments of function, cost, delivery, and robustness. Elegance and consistency come a bad second. It must be easy to change the system in ways that are not predictable or even reasonable—e.g., in response to management directives. At present theorists cannot keep up with this kind of thing, any more than they can with the sheer size and complexity of large software systems.[18]

The report also contains a quote from Christopher Strachey, a computer scientist from Oxford University, from a discussion that was added on the last day to address the gap:

I want to talk about the relationship between theory and practice. This has been, to my mind, one of the unspoken underlying themes of this meeting and has not been properly ventilated. I have heard with great interest the descriptions of the very large program management schemes, and the programs that have been written using these; and also I heard a view expressed last night that the people who were doing this felt that they were invited here like a lot of monkeys to be looked at by the theoreticians. I have also heard people from the more theoretical side who felt that they were equally isolated; they were here but not allowed to say anything. ...

I think we ought to remember somebody or other's law, which amounts to the fact that 95 per cent of everything is rubbish. You shouldn't judge the contributions of computing science to software engineering on the 95 per cent of computing science which is rubbish. You shouldn't judge software engineering, from the high altitude of pure theory, on the 95 per cent of software engineering which is also rubbish. Let's try and look at the good things from the other side and see if we can't in fact make a little bridge.[19]

He is likely referring to Sturgeon's law, coined by the science fiction writer Theodore Sturgeon, which posits that "90% of everything is crap."[20]

Knuth has stated that he feels that at the beginning of the 1970s, academics were good programmers and industry professionals were not. Yet during that decade, as the scope of software that industry wrote increased, the situation reversed itself, and by the end of the decade the academics had drifted out of sync with what was going on in industry and restricted their programming, and therefore their area of expertise, to smaller programs that were no longer useful for generating advice for industry.[21] Basili stated it as, "Researchers solve problems that are solvable, not necessarily ones that are real."[22] The gap, sadly, still persists to this day, despite the occasional bright spot such as design patterns.

But really, there is one obvious cause of the decline in academic research on software engineering. And that cause is me.

Not *just* me, of course. It's me and people like me: the ones who came of age just after the personal computer revolution started in the mid-1970s, which took the move to interactive terminals and accelerated it to light speed. Weinberg commented on this in the silver anniversary edition of *The Psychology of Computer Programming*, observing a team of programmers at a company he was consulting with in the mid-1990s: "More interesting, however, was the coincidence that all of them had learned to program *before* they studied programming formally in school. That's a major change brought about by the personal computer. In my day, I had not even *seen* a computer before I went to work for IBM in 1956."[23]

It's not a coincidence. Beginning in the late 1970s, access to a computer no longer required that you work at a computer hardware company or be associated with a university. Anybody could afford to bring a personal computer home, free from the oversight and advice of more experienced programmers and the methodology of an engineering company, and start programming on their own. And they did just that, in large numbers, learning nothing from the past and reinventing everything, over and over. Both literally and figuratively, we never looked back. Independent software companies thrived while the software divisions of hardware companies shrank, so that the modern software industry was created by people who had never been exposed to engineering rigor.

This is the era in which I grew up as a programmer. I started using computers just as the personal computer was becoming established. But I also majored in computer science at an Ivy League university! As I've said before, all of us were self-taught in how to actually program, and if my professors knew of Weinberg and Mills, they weren't talking about them to the undergraduates. I don't know why it was like this—whether the software world appeared to be changing so fast that this looked obsolete, there was so much else to teach us that there was no room, or maybe they had tried it and been ignored by callow personal computer habitués. Possibly it was seen as not relevant, because these topics tended to get lumped under "psychology" (a term that appears in the title of both Shneiderman's and Weinberg's classic books) or "human factors," which sounds awfully un-engineering-ish.

In 1976, Shneiderman helped found the informal Software Psychology Society, which met monthly to discuss the intersection of computer science and psychology, including software engineering topics. In 1982, the society put on the Human Factors in Computing Systems conference, which led to the creation of the ACM Special Interest Group on Computer-Human Interaction (SIGCHI) conference.[24] Yet for whatever reason, SIGCHI focused much more on the topic of user interaction with computer interfaces rather than programmer interaction with software tools (as eventually did Shneiderman himself).[25] The net effect of all this was that at Princeton, I heard about none of this research, which at the time was only about a decade old and certainly relevant then—and relevant now as well.

I met several people in my early days at Microsoft who had previously worked at hardware companies and left because writing software there felt too slow and bureaucratic: The companies followed the same process for software that they did for hardware. I mean, why not; if you can work unencumbered by established precedent and instead invent everything yourself, wouldn't you want to? No rules! Free money! It's the same pitch that Agile is making today: the less methodology you use, the more you will be free to create works of genius. Unfortunately, rather than trying to learn something about process from these refugees, we enabled them in their quest to be freed from their shackles.

I realize now that this was the difference between Dave Cutler, who was in charge of the Windows NT project when I worked for him early in my Microsoft career, and almost any other executive at the company. Cutler was a generation older than me and had learned the ropes at Digital Equipment, a hardware company. Working there he had acquired an understanding of the need for planning and rigor in software development. Before beginning to write the code for the first version of Windows NT, the team produced a large notebook laying out the internal details of the system, focusing heavily on the APIs provided by each section; a copy of this notebook is now preserved in the Smithsonian Institution.[26] This was before I arrived on the team; I'm not sure what I would have thought of this activity if I had observed it in person. I likely would have wondered why we weren't jumping in and writing code. For that matter, I'm not sure what Bill Gates thought of it (although clearly he allowed Cutler to do it his way). Gates was just young enough that he was able to learn to program on his own, on a terminal like the one that Weinberg accused of leading programmers down the primrose path.

Wilson's "How come I didn't know we knew stuff about things?" moment, described in his SPLASH 2013 keynote, was inspired by his discovering, after ten years in the industry, the 1993 book *Code Complete* by Steve McConnell.[27] This was one of the first books that attempted to assemble wisdom about how to write software. It deserves special mention because it does refer to academic studies to back up its recommendations, at least in areas where studies had been done, such as "What is the right number of lines of code in a single method?" For what it's worth, the consensus on method length from the studies McConnell looked at was that around two hundred lines was getting to be too long.[28] The topic has since been unmoored from any research and now bobs merrily in a sea of impassioned verbiage, such that there are scarcely any positive numbers for which somebody doesn't consider that many lines in a method to be too many. Some people claim that the instant you feel you need a comment in your code, you should instead move that code into a separate method with a sentence-length, camel-cased method name, with said method name serving as the complete documentation; these people walk among us, undetected by the institutions meant to protect a civil society.

Most of the studies cited by McConnell were at least ten years old, since this type of investigation had mostly dried up by then (a second edition of the book, published in 2004, barely unearths any new studies). But at least he referenced them where he could. He even spends five pages discussing Hungarian notation, presenting arguments pro and con without choosing a winner.[29] This is not surprising since Hungarian notation, being a product of industry, has never been formally studied—with both sides instead preferring to continue hurling invectives at each other (in the second edition, he cuts the treatment of Hungarian in half and genericizes it as "standardized prefixes," but he also excises most of the arguments against it, leaving the reader with the impression that it's a good idea).[30]

The IEEE Computer Society, a professional association with similar goals to the ACM, created the Software Engineering Body of Knowledge (SWEBOK), which is summarized in the book *SWEBOK 3.0: Guide to the Software Engineering Body of Knowledge*, known as the *SWEBOK Guide* (the ACM was initially involved in SWEBOK, but pulled out after disagreement on the direction it was taking).[31] This initiative has a cargo cult aspect to it; other engineering disciplines have bodies of knowledge, so maybe if we create one of our own, we will acquire the engineering rigor that they possess. Essentially the IEEE has assembled the current wisdom on software engineering, without passing judgment on the actual value of it.

Given that API design is one of the most critical areas of software engineering (McConnell spends an entire chapter on the subject in *Code Complete*), it is instructive to see what the *SWEBOK Guide* has to say about it. Admittedly the book is less expansive than *Code Complete*, but still it is deflating to find only a quarter of a page devoted to such an important topic. After explaining what an API is, it states that "API design should try to make the API easy to learn and memorize, lead to readable code, be hard to misuse, be easy to extend, be complete, and maintain backward compatibility. As the APIs usually outlast their implementations for a widely used library or framework, it is desired that the API be straightforward and kept stable to facilitate the development and maintenance of the client applications."[32] That's it. This advice is not wrong, although possibly mildly contradictory, but it's woefully incomplete. And what does "should try" mean? Nowhere does it state how to accomplish all these goals or give any references to studies of them.

The *SWEBOK Guide* notes that it doesn't contain detailed information, but points the reader to other literature: "The Body of Knowledge is found in the reference materials themselves."[33] In the case of API design, the redirect is to the book *Documenting Software Architectures*, which is a reasonable book about documenting your software design at various levels of granularity, including down to the individual API layer—yet it is about documenting a design that has been created, not about how to create it in the first place.[34]

Meanwhile, the three-part book *Software Engineering Essentials*, which is meant to provide more detail on SWEBOK and matches it point for point, has this to say about API design, in toto:

An API (application programming interface) is a language and message format used by an application program to communicate with the operating system or some other control program such as a database management system. An API implies that some program module is available in the computer to perform the operation or that it must be linked into the existing program to perform the tasks.[35]

There is no great insight there; it is only a definition of the term, attributed to *PC Magazine Encyclopedia*.

Much of what has been espoused in software engineering in the last twenty years—Agile development, unit testing, the debate about errors versus exceptions, and the benefits of different programming languages—has been presented without any experimental backing. Even object-oriented programming itself has not been subjected to rigorous testing to see if it is better than what came before or just more pleasing to the mind of programmers. As one metareview of the few studies of object-oriented programming put it in 2001, "The weight of the evidence tends to slightly favor OOSD [Object-Oriented Systems Development], although most studies fail to build on a theoretical foundation, many suffer from inadequate experimental designs, and some draw highly questionable conclusions from the evidence."[36]

There are a few stalwart researchers who have continued to do experimental investigations into software engineering. Basili (an IEEE Mills award winner in 2003) deserves special mention, as one of the first and longest practitioners. In addition to a lengthy career as a professor of computer science at the University of Maryland, he spent twenty-five years as director of the Software Engineering Laboratory at NASA's Goddard Space Flight Center. In honor of his sixty-fifth birthday, the book *Foundations of Empirical Software Engineering* was published in 2005, collecting twenty essays from throughout his career.[37] If your curiosity is piqued by titles like "A Controlled Experiment Quantitatively Comparing Software Development Approaches" and "Comparing the Effectiveness of Software Testing Strategies," then I encourage you to learn more about empirical studies. But too often his sort of work winds up in journals like *Empirical Software Engineering* or the *Journal of Systems and Software*, and never crosses over into industry, while working programmers flock to conferences on Agile development and other trendy topics.

When all of us "young blades" banded together in the early 1980s and mounted our successful assault on mainframe computers, we threw the engineering discipline baby out with the mainframe bathwater. The challenge for software engineering is how to get it back.

11 The Future

It has been fifty years since the 1968 NATO conference, when the term *software engineering* entered the vernacular.

The list of concerns voiced at the conference is largely the same one we face today, although progress has been made. The GOTO statement has been demoted from a top-floor suite to the second underbasement. We have better programming languages, even if people are constitutionally resistant to adopting them. Object-oriented programming may not allow us to build programs by placing blocks of code next to each other or make designing usable APIs any easier, but it has given us design patterns, unit testing, and cleaner abstraction between code modules. Programmers still argue over the proper format of variable names and whether tabs are better than spaces, but at least they now do so semi-ironically.

Meanwhile, new approaches to management, while not changing the fundamental task at hand, have acknowledged and adapted to the current reality. Agile, in its various incarnations, has opened people's eyes to the inadequacy of current software project scheduling, and made explicit the need for code that can be understood and modified on an ongoing basis. The open-source movement, in which a team of people who may never have met in person work together to develop software, also emphasizes the need for readable, modifiable code. In addition, since open-source contributors often self-select to work on a project and prove their worth with actual production code rather than interviews, it has highlighted the fact that top-tier university computer science programs don't have a monopoly on producing competent programmers, and established software companies don't have any magic techniques for engineering software.

The most promising prospect for improving software engineering, however, is the move to "the cloud": a company delivering software as a service that runs on that company's computers instead of as software that runs on customer machines.

The period after the IBM PC became the standard platform was the gravy years for the software industry, with companies focused on what became known as *packaged software*. They wrote software, deemed it "good enough to ship" by whatever quality standards they chose to enforce, and then were mostly done with it; all that was left to do was count sales and watch the stock price rise. Customers would report bugs, and companies might provide updated versions, but most of the pain of running the software—acquiring it, installing it, administering it, living with bugs, and rolling out updates—was felt by the customers.

This left programmers relatively isolated from problems arising from their work. If a customer reported a bug, and a programmer could reproduce the bug on their machine, then they could fix it, but otherwise it was easy to treat nonreproducible bugs as flaky problems that could be ignored. Occasionally a bug would be widely reported enough that programmers felt angst over it (for example, the Zune "Day 366" bug that messed up New Year's Eve 2008 party plans, which was covered in the mainstream press), but mostly they could enjoy a blameless sleep and try to tackle the bug on their own schedule. Some teams at Microsoft even had a separate group of engineers responsible for fixing bugs in software once it had shipped, so somebody else had to deal with bugs while the original author was off cranking out cool new stuff; notably, but not surprisingly, this "sustained engineering" role, like the tester role, was viewed as lower on the engineering totem pole than developers.

With the shift to software running in the cloud—such software is generically called a *service*—all that changed. The company that writes the software also installs it and keeps it running; the customer accesses it via a web browser. Forget about suggesting that customers try to turn the machine off and then back on again; the end users have no control at all over software running on a machine in a faraway data center. They have to report all problems to the company—and to keep customers happy, it is best if your software has already figured out that it is malfunctioning, so you can be alerted before customers notice. Furthermore, it is unlikely that you can stop a machine for an hour while you debug the problem. Issues have to be figured out from telemetry data that is continuously captured on the running system—recording information about the system, logging what files are accessed, tracking database queries, and so on.

When I worked on Windows NT in the early 1990s, every night before leaving work we would start "stress tests" running on all our computers. These automated tests were a mix of programs that ran continuously, performing basic operations such as reading and writing files, or drawing graphics on the screen; the goal was to see if Windows NT could make it through the night without crashing or hanging.[1] Developers grew to dread the morning e-mail indicating that a machine had failed in code that they owned. By general agreement, stress failures needed to be investigated before the machine could be reclaimed, since it might be the unique example of an intermittent bug. Frequently the problem had hit on another developer's primary machine, leaving them dead in the water until the investigation was completed. I recall those debugging sessions being the most stressful, because somebody else was stuck until I was done. And forget about repro steps; the bug likely hit due to precise interactions between various pieces of software, which might never reoccur again, so you had to do what you could to retroactively figure out what went wrong, by forensic digging into the current state of memory to try to isolate the first fault and thence the code defect. How exhaustively you investigated before giving up on any given failure was up to you, but the pressure ratcheted up if it happened again; woe betide the developer who continued to be unable to fix a recurring stress failure.

On a service, every investigation is like that. Suddenly, bugs become an immediate concern for developers. A widely used strategy is to give the developers pagers and put them on call to look into issues on a rotating basis, transforming developers into living exceptions handlers. This forces programmers to directly feel the pain of bugs, and it also inspires them to make sure the monitoring and alerting is accurate; nobody wants to miss a real issue, but nobody wants to be woken up in the middle of the night by a false alarm. And if your telemetry data isn't rich enough to debug the failures, then you have lots of motivation to improve it, and quickly.

On the plus side, because the bugs are happening on machines that companies own, it is easier to keep track of them and figure out which ones cause the worst disruptions, and then go back and figure out how the problem could have been avoided—and how future problems of the same sort could be avoided. Was it poor design choices? Was it an API with undocumented side effects? Was it a unit test that should have been written? Was it a test that was written, but not run at the right time? Was it a step on a deployment checklist that was missed because somebody was not being careful or it wasn't written down? It is now much more obvious if people are skipping out on these engineering-like behaviors that programmers had previously been able to ignore.

At some point, you could even conceive of tying this back to the so-called religious debates that have befuddled programmers through the years. Do you prefer your particular coding style? Do you think your favorite language leads to fewer bugs or more rapid development? Do you like Hungarian prefixes on your variable names? It's unlikely that one company is going to produce enough data to individually come up with an answer, but if you analyze this across the industry, it's possible that actual answers could emerge from the murk. Even if companies are not able to answer the questions themselves, they would at least be motivated to care about the answers (and sometimes to impose whatever mitigation was discovered onto their teams) because services have a nice way of converting these decisions into dollar amounts. If you have more bugs than another company that has made a different choice of coding style, if your service runs more slowly, if it is harder to deploy ... that all translates to more money you need to spend somewhere to keep it running, which makes it harder to compete in the market.

Essentially, running a service imposes an automatic "no bullshit" filter on the old wives' tales that permeate software engineering.

As an added benefit, the way data is transferred between components of a service, or between a client machine and service in the cloud, is much closer to the ideal of object-oriented programming than is typically achieved with software running on a single computer. In the early days of computer networking, when machines were all on the same local area network, the network was low bandwidth but also low latency; sending a lot of data through the network was slow, but a small network packet could reach the other machine fairly quickly. As a result, early network protocols focus on packing as much data into as little space as possible, which made writing the code to handle them tricky; it was processing this sort of incoming network packet that led to the Heartbleed worm in 2014.

Modern networks connecting a client to the cloud have much higher bandwidth. The computer network at Microsoft back in 1990 ran at 10 megabits per second shared by the entire floor of a building, while today it is typical for a home broadband connection to support 100 megabits per second directly to a single house, with the backbone connections that get you to the Internet running much faster than that. The connections have a higher latency, however, since a packet has to make multiple hops between computers. As a result, the amount of data in a packet doesn't matter as much—a large packet doesn't take noticeably longer than a short one to reach its destination—and communication is usually encoded in a more verbose format known as Extensible Markup Language (XML).

XML is a slightly more categorized version of the text strings used to communicate between pipelined UNIX commands, which previously had been the most effective implementation of "object-oriented" software building blocks. In older network protocols containing tightly packed binary data, the meaning of any given byte depended on its exact location in the packet, and it was hard for a human to parse the data to figure out what it meant; entire separate programs known as *packet sniffers* existed to assist in debugging network traffic (the sniffers we used in my early days at Microsoft were entirely separate machines, quite expensive to replace if you happened to drop a network cable into the back of one while it was turned on—not that I would know anything about that). XML, by contrast, uses human-readable tags to identify data, so a programmer can scan the XML request itself to figure out its meaning, and code to parse XML can be written more generically and therefore more safely (Microsoft Office uses XML in its newer file formats to avoid the problem of exploits hidden inside the old binary format). It's similar to the difference between reading raw bytes of machine language and reading code in a higher-level language.

A second benefit is that XML communication is much more forgiving of version mismatches than earlier protocols. In particular, if a client sends data in a request that the server doesn't expect, it can simply ignore it, which makes it easier to expand a network protocol without breaking interoperability with older clients. If early network protocols were like the tightly bound API calls in a performance-focused language like C++, where a single added API parameter could break the code unless all callers were also updated, XML is more like the Smalltalk messages-with-named-parameters approach—a little slower, but also less fragile in the face of changes.

That said, there is an interesting bifurcation going on. At the same time that programmers are working on cloud services with almost infinite computing power, they are writing programs to run on smaller devices. Just as a generation of programmers came of age on the resource-limited mainframes and minicomputers of the 1960s, and a new generation came of age on the resource-limited personal computers of the 1980s, it's possible that another generation will learn to program on resource-limited phones and tablets. What remains to be seen is whether the Third Great Software Resource Bottleneck will teach programmers all the same bad habits of the previous generations, favoring performance and convenience, and less typing of course, at the expense of clean design.

The services approach can be difficult for programmers experienced in packaged software to adapt to. Services require a mind shift away from performance toward maintainability and reliability, and a further mind shift away from concentrating on shipping software toward focusing on keeping it running. Microsoft famously has a Ship-It Award where everybody who contributes to a project that ships to customers receives a small commemorative metal sticker to attach to a plaque, with the resulting accretion forming a record of each employee's time at Microsoft.[2] The award was created in the early 1990s after a few large projects visibly flamed out and were canceled, with nothing to show for years of work; it was meant to emphasize the importance of actually finishing software and delivering it to customers. Despite a legendary stumble at the launch of the program,[3] people do care about their Ship-It Awards, and many Microsoft employees proudly display their plaques in their offices.

Flash-forward fifteen years and your humble narrator, as part of the duties managing my team in Engineering Excellence, was solely responsible for determining whether a piece of software written at Microsoft was significant enough to deserve a Ship-It Award. The process by which this authority came to be vested in my position was somewhat convoluted, but I did my best to dispense justice as I saw fit (new versions of Office and Windows: yes; update to the point-of-sale software used in the Microsoft Store: no). The Ship-It program had been designed for packaged software, but around this time the large services teams were realizing that shipping is only the beginning; the real trick is to keep a service running. They began agitating for an equivalent recognition of this to adorn their plaques (the working title was Run-It Award), yet in the end their request was shot down—a legacy of the "it's just sustained engineering" mentality.

In 2009, a lot of programmers at Microsoft were beginning to work on services, so we in Engineering Excellence went around the company asking those who had already been through the transition for their advice (besides "give us a Run-It Award"). The most insightful comment came from an engineer on the Exchange e-mail service who had previously worked on packaged software. After walking through a list of all the things he had needed to relearn in order to be effective working on a service, he ruefully said, "The problem is, if I went back in time two years and told myself all of that, I wouldn't have believed it." Brooks at one point quotes Benjamin Franklin, writing in *Poor Richard's Almanac*: "Experience is a dear teacher, but fools will learn at no other."[4]

The move to services is a step in the right direction, but more needs to be done.

Gawande's book *The Checklist Manifesto* is subtitled *How to Get Things Right*. At one point he talks to a structural engineer about how the construction industry has changed:

For most of modern history, he explained, going back to medieval times, the dominant way people put up buildings was by going out and hiring Master Builders who designed them, engineered them, and oversaw construction from start to finish. ... But by the middle of the twentieth century the Master Builders were dead and gone. The variety and sophistication of advancements in every stage of the construction process had overwhelmed the ability of any individual to master them.[5]

Today's software engineers are master builders, and they are in danger of being overwhelmed.

Gawande also discusses the hero test pilots from the early days of high-speed aircraft, as documented in Tom Wolfe's book *The Right Stuff*. They succeeded through improvisation and daring, but eventually "values of safety and conscientiousness prevailed, and the rock star status of the test pilots was gone."[6] The universe applies a natural corrective to test pilots who fail to adapt, but no such effect exists in software; the middle and upper management of software companies are full of successful self-taught programmers. Hopefully, more recent graduates have heard about design patterns and unit tests; while those are a long way from providing a scientific basis for software engineering, at least they have opened people's eyes to the fact that accepted knowledge is possible in the software industry, which in turn should make them more amenable to learning new ways if we can discover them. But of course they do have to be discovered first.

Otherwise, we will have to confront another quote from Gawande's book:

For nearly all of history, people's lives have been governed primarily by ignorance. ... Failures of ignorance we can forgive. If the knowledge of the best thing to do in a given situation does not exist, we are happy to have people simply make their best effort. But if the knowledge exists and is not applied correctly, it is difficult not to be infuriated. ... It is not for nothing that philosophers gave these failures so unmerciful a name—*ineptitude*.[7]

Are programmers inept? I don't think they are in the exact way Gawande indicts medicine, the prime focus of his book. He is talking about situations in which the knowledge of how to deliver proper medical care is unquestionably available and agreed on by all involved, but for various reasons is not applied properly. Gawande sees parallels to this struggle in "almost any endeavor requiring mastery of complexity and of large amounts of knowledge"—a list that to him includes foreign intelligence failures, tottering banks, and (little does he know) flawed software design.[8]

The argument against labeling programmers inept is that they have not even reached the stage of knowing the right way to do their jobs. That's because they have stopped trying to figure this out, which isn't a great defense: we are still inept, just in a slightly different way.

In 1986, Brooks wrote an essay titled, "No Silver Bullet"—a reference to the ammunition reputedly needed to kill werewolves. "We hear desperate cries for a silver bullet, something to make software costs drop as rapidly as computer hardware costs do," he lamented,

But, as we look to the horizon of a decade hence, we see no silver bullet. There is no single development, in either technology or management technique, which by itself promises even one order of magnitude improvement in productivity, in reliability, in simplicity. ... Not only are there no silver bullets now in view, the very nature of software makes it unlikely that there will be any.[9]

Brooks describes what he feels is the *essence* of the problem, the inherent difficulties of software: complexity, conformity (the need to fit new code into an existing API), changeability, and invisibility (the difficulty of visualizing the internals). He lists various developments that may help, including object-oriented programming (which at that time was just beginning its transformation into Object-Oriented Silver Bullet Amalgamated Holdings), but doubts that any of them will supply the magic needed.

Nine years later, almost at the end of the time period he was considering in the original paper, Brooks produced a follow-up essay "'No Silver Bullet' Refired," discussing reactions to the original. He wrote, "Most of these attack the central argument that there is no magical solution, and my clear opinion that there cannot be one. Most agree with the arguments in 'NSB,' but then go on to assert that *there is indeed a silver bullet for the software beast, which the author has invented.*[10] The italics are mine, because I can think of no more succinct summary of the ongoing procession of alleged software panaceas.

Brooks ends his second essay with a short section titled "Net on Bullets—Position Unchanged." He quotes noted software observer Robert Glass, in an essay in *System Development* magazine: "At last, we can focus on something a little more viable than pie in the sky. Now, perhaps, we can get on with the incremental improvements to software productivity that are possible, rather than waiting for the breakthroughs that are not likely to come."[11]

The history of software engineering has been a search for the silver bullet. Structured programming, formal testing, object-oriented languages, design patterns, Agile methodologies—all are useful, but none alone can slay the werewolf. I personally lived through all these transitions, even structured programming; due to spending my early years in a self-taught bubble of Fortran and BASIC, I was able to experience the twilight of GOTOs firsthand, ten years later than the rest of the industry. Each of these started out small, gained supporters, and wound up being hyped as the solution to all programming problems, leading to inevitable disappointment and disillusionment. I can understand why programmers hope beyond hope for the silver bullet and have eagerly glommed onto a succession of shiny objects. "Any port in a storm," as the saying goes. Unfortunately, as Parnas put it, "[Programmers] have been fed so many 'silver bullets' that they don't believe anything anymore."[12]

Moreover, the software industry has gotten in the habit of abandoning old silver bullets once they get tarnished. It's a binary view of the world: a programming technique is either going to solve every problem or it's useless. When Microsoft began hiring software test engineers to test the software, the developers, who previously had been responsible for this, quickly segued into "throw it over the wall to test" mode. In the mid-2000s, Microsoft replaced software test engineers with software design engineers in test, responsible for writing automated tests, and stopped relying on manual testing. With the later move to unit tests, which are written by developers, the company got rid of a lot of the software design engineers in test and the user-interface-level tests that they provided. Now with the transition to the cloud, there is an emphasis on "test in production," where updates to the software are quickly deployed to a small percentage of real customers, with checks in place to quickly detect problems and roll back the update. Each new technique is useful, but it should be viewed as another arrow in the quiver, not the be-all and end-all that obviates the need for what went before.

In 1978, Harlan Mills predicted that "The next generation of programmers will be much more competent than the first ones. They will have to be. Just as it was easier to get into college in the 'good old days,' it was also easier to get by as a programmer in the 'good old days.' For this new generation, a programmer will need to be capable of a level of precision and productivity never dreamed of before."[13] As with almost everything Mills wrote, this is as true today as it was back then.

So what is to be done?

Most of my recommendations involve changing how students are taught software engineering in universities. I am not entirely blaming universities; one of the main reasons they do not know what to teach is because industry is too self-satisfied and complacent to interact with academia. If you ask companies what they would like colleges to teach, they will likely start talking about so-called soft skills: communication, being on time, and working well with others. That's nice, but it's not surprising that a bunch of adults, who now have families, mortgages, and other responsibilities, are able to recognize that college seniors aren't quite as mature as they are. It's more difficult to determine what technical skills may be lacking in college graduates, particularly when those same skills may be lacking in the experienced employees as well.

And in the spirit of *The Checklist Manifesto*, I am not going to be too prescriptive. As long as industry and academia have talked and agreed on what is to be done, I will trust their judgment. But by "agreed," I don't mean the ACM has a conference where professors and industry researchers show up; there has been enough of that already. The ACM, to its credit, has over thirty special interest groups, including ones focused on computer science education, programming languages, and software engineering, but they do not get enough traction among working software professionals. I want software engineering programs to change their curriculum, and companies to care.

In any case, it's a question of timing: future programmers usually are exposed to software education before they are exposed to software jobs, so let's begin there. Through a combination of academia, industry, and divine providence, the following areas need to be addressed.

Force Students to Learn Something New

When the independent software industry emerged in the 1980s, it reinvented everything about managing software projects that had been written in the 1960s and 1970s. It is hard to fathom how much time and energy was wasted because of this. Yet there is no reason why it won't happen again with the next generation of programmers.

Some students arrive in college with no knowledge of how to program, but many have significant experience. The sooner students are disabused of the notion that they know everything they need to know, the sooner they will become receptive to new ideas. I made it through my Princeton computer science classes, and to some extent my early years at Microsoft, on the basis of the skills I had taught myself in high school, writing BASIC games for the IBM PC. The classes weren't easy; I had to write a lot of code and spent a lot of late nights in von Neumann getting my programs to work. But this did nothing to open my mind to the wide world of software engineering.

Forcing students to learn something new rather than get by on their existing skills will make them humble. And as jazz trumpeter Wynton Marsalis said, "The humble improve."[14] One example is the details of different algorithms for sorting arrays, which have names like *bubble sort* and *selection sort*. Different sorting algorithms perform better with certain data, which Knuth (in his landmark series of books *The Art of Computer Programming*) and others worked out a long time ago. When I asked programmers to write a sort algorithm on the whiteboard during a Microsoft interview, I didn't care how they did it or if they knew the name of the algorithm. I was looking for that hard-to-capture ability to write code and get it to work, summarized in the statement, "They can code." But why not expect that students can rattle off the intricacies of different sorting algorithms? Ironically, information floating around the Internet claims that Microsoft is looking for this level of detail during an interview, but I never saw it, although if you look over the slides from Stanford University's class Problem-Solving for the CS Technical Interview or scan discussion threads on the website Reddit, it seems that some companies are.[15]

Even if students don't remember all the details of algorithms, at least they will take away the awareness that there is a lot of knowledge out there, available when they need it. There are situations where your sorting algorithm does matter a lot, and you want students to develop the instinct of recognizing them.

While you're at it, have students do up-front estimates of how much time they think it will take them to complete various phases of a project and then look back to compare that to how long they actually took. These will be smaller projects, so the estimates won't be as far off as in the real world, but it can't hurt to show them that even for smaller projects, estimating software is a perilous endeavor.

Certainly a new graduate may encounter a situation where deep engineering knowledge is not needed. Coding camps, which claim to turn anybody into a programmer in a few months, have a mixed reputation, but they do indicate that the basic knowledge required for certain programming jobs can be taught relatively quickly.

If students wind up working in an Agile Eden, on a small team that stays together for a long period of time, with a single customer who is engaged in overseeing their software, calling well-documented APIs in a well-understood environment—then great, they can dial back their application of software engineering principles and relive their halcyon days. But if not, they need to have the core engineering knowledge. It's a lot easier to know the underlying principles and choose not to apply them than it is to not know them and be in over your head.

Students should also study their history. In June 2001, a conference was held in Bonn, Germany, featuring presentations by sixteen influential software pioneers; the results (including videos) were collected in the book *Software Pioneers*.[16] It's an impressive list: Wirth and Brooks; Dahl and Nygaard, who designed Simula; Friedrich Bauer, inventor of the stack; Kay, who came up with Smalltalk and the graphical user interface; Hoare, who did early work on program correctness, and his fellow Algol 68 denouncer Dijkstra; Parnas, who wrote some of the earliest papers on modularizing software; and others whom I have not mentioned only due to lack of room in this book.

The fact is, these luminaries won't be around forever. Dijkstra, Dahl, and Nygaard all died the following year, within six weeks of each other; Bauer died in 2015. John Backus, inventor of Fortran, who was invited but could not attend, died in 2007. It would be a tremendous disservice to them if all the programmers who are benefiting from their insight did not acknowledge their contributions, and a tremendous waste of time if their wisdom was not dispensed to the next generation of programmers.

Work to Level the Playing Field

We need programmers; we can't afford to exclude anybody before they leave the starting gate. You want to attract anybody who is interested in programming, high school geek or not, whatever their race and gender. Even if the sum total of what motivates a person to major in computer science is a single Hour of Code tutorial or the fact that they like to play video games, they should still feel welcome. Hopefully this will produce a double benefit: making other new programmers feel more welcome at school, and producing software engineers who are more teachable and open to lifelong learning. The alleged rise of the "brogrammer," that annoying fraternity kid now reborn as a web developer, is in some ways an encouraging sign of software careers opening up to a wider audience.

Some universities have split their introductory classes into multiple tracks, so that students who have no experience programming are not intimidated by the propeller-heads.[17] There is also a push to organize first-year programming classes around projects that are more interesting to students than pure algorithmic noodling, such as robotics or game programming.[18]

Another tactic is ensuring that having been in the computer club in high school is unlikely to provide a big advantage. Carnegie Mellon University teaches a language called ML in its introductory class. ML is an elegant but nonmainstream language, belonging to a class of languages known as functional languages—that is, not a procedural or object-oriented language. It's something that programmers may never use in their professional career (although not surprisingly some programmers today claim that functional programming is, at last, the silver bullet that will cure all programming ills); the big benefit of using ML in an introductory class is the likelihood that none of the incoming students has used it in high school (they are unlikely to have learned *any* functional languages), so it places everybody on an equal footing at the start.

Related to this, there has been a fair bit of discussion about how to increase the percentage of women majoring in computer science, sometimes expanded to include having underrepresented minorities reach proportional representation. Harvey Mudd College has gotten a lot of press recently for reaching a proportion of almost 50 percent women in its computer science program, with a multitrack approach to introductory classes being one of its main strategies. Carnegie Mellon also has close to 50 percent women in its computer science program.

Computer science enrollment waxes and wanes, depending on whether the news is focusing on stock option millionaires or tech companies going bankrupt. A 2017 report by the Computing Research Association shows that since 2006, when computer science enrollment was at a low point following the bursting of the dot-com bubble in the early 2000s, the number of computer science majors has tripled, reaching an all-time high that is almost twice the number at the peak of the dot-com boom.[19]

The percentage of female computer science majors was 14 percent in 2006, dropped to 11 percent in 2009, and had climbed back to 16 percent by 2015, the last year covered by the report. Meanwhile, the percentage of underrepresented minorities hovered around 10 percent the entire time, ticking up slightly to 13 percent in 2015.[20] Given the overall rise in enrollment, these still represent large increases in the absolute numbers of women and underrepresented minorities majoring in computer science, but they remain much lower percentages than in the overall population (those underrepresented minorities, as defined for the report, constituted around 30 percent of the US population in the 2010 census).[21]

Female programmers have a secret weapon here. Both academia and industry need to do more to connect students and employees with the IEEE and ACM; these are the relevant professional societies, and people working in software engineering should know and care about them. The secret weapon for women is an annual conference known as the Grace Hopper Celebration of Women in Computing, named after the author of the first compiler who was one of the guiding forces behind COBOL (which, despite being the butt of jokes today, was a big step forward in its time). The conference has both career-related and technical talks. Such material is available in a lot of places, but the key to the Grace Hopper Celebration is that people, both students and professionals, attend it in large numbers; it is the conference that Microsoft, by a large margin, was the most interested in sending employees to. (The conference does allow men to attend and in fact actively encourages it.)

Teach Students to Work with Larger Pieces of Software

Back in 1980, Mills wrote,

It is characteristic in software engineering that the problems to be solved by advanced practitioners require sustained efforts over months or years from many people, often in the tens or hundreds. This kind of mass problem-solving effort requires a radically different kind of precision and scope in techniques than are required for individual problem solvers. If the precision and scope are not gained in university education, it is difficult to acquire them later, no matter how well motivated or adept a person might be at individual, intuitive approaches to problem solving.[22]

It is common nowadays for students to do class programming projects in teams of two to four, to expose them to the issues that arise when you are not the only person working on a program. Even back in my day, I did several class projects working with a partner. This is a laudable attempt to expose students to some of the situations they will encounter working in industry.

Unfortunately, working with two to four people for one semester is not a large enough step up from a one-person project. Yes, you will see somebody else's coding style, have to divide up the code and work out an API between each person's contribution, and practice your interpersonal skills—all valid and possibly eye-opening experiences. But fundamentally, you can engineer a project like that using the programming skills you have taught yourself; if the API interface you are calling was worked out with a fellow student, it is unlikely that there will be confusion about the functionality, and in a semester you are not going to forget how the code works, so future maintainability won't be an issue. Recalling Brooks's description of the difference between a program and a programming system product—requiring both a move from a single author to multiple authors, and a move from a single component to a program that is built up from connections across API boundaries—working on small group projects gives you a slight nudge in both dimensions, but you still remain firmly in the starting quadrant.[23]

Larger pieces of software will also let students practice reading code, and teach them the benefits of clear variable names, code comments, and the like. To get the benefits here, they need software that is much more complicated than what students could create on their own. It's not like civil engineering, where if you can model the stress on one steel beam, you can extrapolate to an entire bridge. Perhaps some companies with large codebases would volunteer their code, viewing it as an opportunity to get name recognition among potential future employees. Alternatively, if companies won't fork over their code, there are large open-source projects that would provide fruitful training grounds for students too. Care should also be taken to choose code written in a variety of languages, to help students get a better understanding of what languages are most suitable to what sorts of problems.

Debugging is also vastly easier on small programs, especially since they tend to run on small amounts of data. In college, I made do with what was termed *printf debugging*, after the printf() API in C that is used to print to the console; if my program was not working as expected, I would litter it with temporary printf() statements to display the contents of my variables at different points and then look through the output trying to find the point of the first fault. This works great on college-scale projects, but falls apart rapidly in the corporate world—at least for most of us. Thompson, who designed and implemented the first UNIX system, has stated that he much prefers printf debugging.[24] But as Baird put it, "If you are a genius like Ken Thompson, you are going to write good code."[25] The rest of us need to move beyond printf debugging to get our code working, as I rapidly discovered once I started working on the internals of Windows NT at Microsoft (the fundamental skill is learning to use a specialized piece of software called a *debugger*, which can examine the memory of another program).

Debugging can in ways be compared to a doctor diagnosing a patient—a skill that medical schools certainly spend a lot of time teaching. Of course, the doctor doesn't first have to learn how to build a human being, which leaves more room for other topics, but debugging is rarely taught to students. There are tools and techniques that can be used for debugging large programs that could be taught in college if large enough codebases were used to practice on, but I arrived in industry completely unaware of them.

Parnas expressed hope that internships would expose students to these sorts of issues.[26] The problem is that internships are primarily extended job interviews as well as advertisements for the company. For both reasons, companies tend to smooth the path for interns and avoid exposing them to too much code, no matter how well organized it might be. Requiring interns to assimilate a large body of code would be discouraging and also soak up a significant percentage of the internship. But they could certainly be required to do it for a college course.

Emphasize Writing Readable Code

Code is usually written once and read many times, but historically the emphasis has been on making the writing easier, at the expense of the reading. Requiring that students participate in inspections—formal inspections, not ad hoc code reviews—would be extremely valuable for everyone, not necessarily to find defects, but to ensure that students were writing readable code because they would need to read each other's code in order to participate in the inspections. Inspections can be daunting not only for the person whose code is being inspected (who can feel like they are suffering through an inquisition) but also for the inspectors (who can feel pressure to find issues). But these can be mitigated, partly by having an experienced person leading the inspection, and partly by having everybody in turn volunteer code to be inspected. And if students complain to their professors that inspections are unfair because they haven't been taught what their code *should* look like—well, that would be a great forcing function for professors to go figure that out.

There are other techniques that have attempted to produce more readable code. In 1984, Knuth published the paper "Literate Programming" about writing programs with a primary focus on making them comprehensible to others. He enthused, "The impact of this new approach on my own style has been profound, and my excitement has continued unabated for more than two years. I enjoy the new methodology so much that it is hard for me to refrain from going back to every program that I've ever written and recasting it in 'literate' form."[27]

Literate programming involves the programmer authoring a file that contains both the code and an explanation of it, mixed together, with the explanation as the primary focus, and the code appearing in small fragments following the relevant explanation; fragments of code can reference other fragments of code, similar to an API call. A separate program then parses this original file to produce two pieces of output: a file containing all the fragments pulled together so they can be compiled, and a formatted document containing all of the explanations of the fragments (Knuth named the first implementation of this system "WEB," explaining—this was in 1984, remember—"I chose the name WEB partly because it was one of the few three-letter words of English that hadn't already been applied to computers").[28] When my kids took drivers' education, they were taught to do "commentary driving," in which they explain their thoughts out loud as they are driving. Literate programming is sort of like that for coding, except you write your commentary down alongside the source code.

Knuth did an experiment with seven undergraduates one summer, teaching them literate programming; six of them loved it because, as he put it, "it blended well with their psyche."[29] The methodology doesn't have broad adoption; Knuth commented that if one in fifty people are good at programming, and one in fifty people are good at writing, it is hard to find people who are good at both. Knuth successfully used it for the popular text-formatting program known as T_EX, although for much of it he was the only author.

Literate programming's greatest success may be a computer graphics renderer—software that takes a 3-D representation and turns it into an image—written by Matt Pharr and Greg Humphreys, which they documented in the book *Physically Based Rendering*.[30] They were inspired by the book *A Retargetable C Compiler*, which was also created using literate programming (in fact, as a collaboration between a Bell Labs researcher, Christopher Fraser, and a Princeton professor, David Hanson).[31] In both cases, the actual book was generated from the original file that had the code fragments mixed in; the book therefore served as complete documentation of the algorithms used for the program and was easy to keep in sync with the code.

Pharr and Humphreys call literate programming "a new programming methodology based on the simple but revolutionary idea that *programs should be written more for people's consumption than for computers' consumption*."[32] Along with their Stanford professor Pat Hanrahan, they won a Scientific and Technical Academy Award for the book; if you watch videos of the presentation, not only do you see Kristen Bell and Michael B. Jordan do a surprisingly credible job of explaining recent advances in rendering technology, you get to hear Knuth be thanked in an Oscar acceptance speech.

Is literate programming a great idea? I honestly don't know; even if Bell does have a copy of *Physically Based Rendering* on her Kindle, as she jokingly claimed, it's just one example of a successful program. Knuth himself wrote, "My enthusiasm is so great that I must warn the reader to discount much of what I shall say as the ravings of a fanatic who thinks he has just seen a great light."[33] In the foreword to *Physically Based Rendering*, Hanrahan observes, "Writing a literate program is a *lot* more work than writing a normal program. After all, who ever documents their programs in the first place!? Moreover, who documents them in a pedagogical style that is easy to understand? And finally, who ever provides commentary on the theory and design issues behind the code as they write the documentation?"[34] Maybe it is more work—but maybe that is work that programmers need to do.

Literate programming is an interesting example of how hard it is for new ideas in programming to emerge from academia. It was invented by Knuth, who has impeccable credentials and broad name recognition; it has an example in T_EX that is widely used and unusually large for a piece of software produced by an academic; and it had positive results from the informal study that Knuth performed. Nonetheless, it has a hard time getting airplay when competing with trendier initiatives such as Agile, whose proponents' livelihoods are based on its success.

Whether it is a literate program or something else, if the provider of an API created more formal documentation on what the API did, rather than just the method name and parameter list, it would mitigate API confusion. In Java, a method has to list every exception that it throws as well as every exception thrown by a method it calls that it does not itself catch—in other words, the complete set of exceptions that a caller of this method can expect to have thrown.[35] This requires more thinking and typing by the programmer, but it does clarify some of the potential side effects of the method, allowing the calling code to be more robust, since unclear side effects inside an API are a source of errors. Parnas, in a 1994 paper with Jan Madey and Michal Iglewski, proposed a symbolic notation for specifying the side effects of API calls.[36] Such things make programmers blanch, but if it helps clear up confusion, we need to consider biting the bullet and having programmers learn to read the notation.

Relocate Certain Well-Understood Topics

The university majors computer science and software engineering are used inconsistently although often interchangeably. But certainly, you have people who undertake a course of study in college that is more theoretical than what they apply on the job. As McConnell wrote about the issue of computer science majors becoming software engineering professionals, "This puts computer scientists into a technological no-man's land. They are called scientists, but they are performing job functions that are tradition-ally performed by engineers without the benefit of engineering training."[37]

We need a real software engineering major that focuses on the engineer-ing practices. Yet undergraduate curriculums are already jam-packed. How will they make room to teach more fundamentals?

One answer is to push a few time-honored subjects out of the software engineering major. There are areas of computer science where the theory and practice have been fleshed out reasonably well over the years, includ-ing graphics, compilers, and databases. These are often taught in under-graduate courses, but the reality is that unless a student goes on to work in that specific area, these courses don't add much value, except as a chance to write more code. And if students do wind up working in that area, they can look up that knowledge. The problems with software engineering are generally not in finding the correct algorithm; they involve translating that algorithm into code that works correctly. So while it's good for all students to learn fundamental algorithms for sorting and such, the more advanced ones aren't necessary.

Of course, these are frequently the areas that professors specialize in, so teaching such classes has appeal to them. I'm not saying they shouldn't be taught at all; I'm saying they should be moved over to a real computer sci-ence major that does not claim to teach software engineering or, instead, offered to graduate students.

Pushing these topics to graduate school would allow a student who was interested to concentrate on them, which would allow students to differ-entiate themselves more. Currently, software engineers coming out of col-lege are viewed as fungible; it is expected that any programmer, if found competent by whatever hiring procedure is used, can go work on any part of a program. As software becomes more and more complicated, however, it makes more sense for people to specialize in different areas.

Mills wrote about specialization on a surgical team:

A surgical team represents a good example of work structuring, with different roles predefined by the profession and previous education. Surgery, anesthesiology, ra-diology, nursing, etc. are dimensions of work structuring in a surgical team. The communication between these roles is crisp and clean—with a low bandwidth at their interface, e.g., at the "sponge and scalpel" level, not the whole bandwidth of medical knowledge.[38]

Gawande also talks about specialization in his book on checklists. Following his guidance, the way to ensure that software was secure, fast, reliable, or whatever other aspect you were concerned about, would not be to make a long checklist that had to be followed by every programmer. It would be to hire programmers who were specialized in security, performance, reliability, or whatever, and then have a checklist item that simply read, "Have you checked with the security/performance/reliability/whatever expert that they are happy with this software?" In an undergraduate education, not all these areas can be covered. Making them graduate specialties would solve that problem and give a graduate degree higher status in industry. Companies, in turn, would learn to favor them when hiring people to fill these specialized roles.

Pay Attention to Empirical Studies

In 1995, Brooks wrote, "In preparing my retrospective and update of *The Mythical Man-Month*, I was struck by how few of the propositions asserted in it have been critiqued, proven, or disproven by ongoing software engineering research and experience."[39]

Empirical studies never went away; the Empirical Studies of Programmers workshops continued, and the journal *Empirical Software Engineering* is still published. A well-known (within the field of empirical research) conference was held at Dagstuhl Castle in Germany in 2006. The book *Making Software*, from 2011, has thirty essays on empirical software engineering, including "How Effective Is Test-Driven Development" and "How Usable Are Your APIs."[40] The Psychology of Programming Interest Group in England continues to put on an annual conference.

What changed was the appetite of programmers for consuming this material. With the PC explosion in the late 1970s and 1980s came the emergence of an independent software industry, filled with self-taught programmers who quickly realized they could make lots of money by continuing the practices they had taught themselves. They felt no need to look at research on how to improve, so they didn't. Whatever methodology turned out to be critical for their success, such as tracking bugs or scheduling, they gradually and painfully reinvented, unaware that the prior generation had already written extensively on how to address these problems.

The irony is that programmers generally like to get uncommon insight into the way the world works. In 1984, when I was a senior in high school, I attended a weeklong camp at the University of Waterloo for students who had done well in an annual math competition. At this meeting, essentially the annual gathering of the Future Programmers of Canada Society, a fellow attendee was reading a book by a baseball writer I had never heard of named Bill James. The book was all about his attempts to bring mathematical rigor to the game: taking long-unquestioned "rules" of baseball, such as using wins as a measure of pitcher quality or the value of a stolen base, and trying to answer them by mining the rich trove of statistical data that had been built up over the previous century. I immediately loved this idea and became a James fan, as many other programmers are. But somehow this never translated into thinking about whether the long-unquestioned "rules" of software could also be studied and analyzed.

Glass, in his collection of essays *Software Conflict*, asks, "So why don't we have an experimental component to our field's science and our field's engineering?" and continues,

There are two reasons that I can think of:

1. Experiments that are properly controlled and conducted are hard and expensive to conduct. It's not enough to rope three undergraduate students into writing 50 lines of Basic and then compare notes. If an experiment is going to be meaningful, it ought to involve real software developers solving real software problems in a carefully pre-defined and measured setting.
2. The engineers and scientists in our field are neither motivated nor prepared to conduct meaningful experiments. Advocacy [by which Glass means writing that promotes a methodology without any supporting experimental evidence] has been with us for so long that is just doesn't seem to occur to anyone that there's a component missing from our research. And without motivation to supply the missing component, no one is getting the proper intellectual tools to know how to conduct experimental research.

Glass does allow, "Perhaps the words *no one* in the preceding paragraph is an example of going too far. For example, the folks working on the intersection of software and psychology, the 'empirical studies' folks, are doing fairly interesting experimental work."[41]

Just because empirical studies of programming are hard doesn't mean that researchers should not be doing them—and it certainly isn't an excuse for companies to ignore what has already been studied. The name *empirical studies* and alternative term *software psychology* need an upgrade. What we are talking about is *programming science*. Normally I would put this one on academia and say that it needed to do more of this sort of research. But industry is also not conditioned to care about this sort of research because it has been so successful without it.

When I worked at Microsoft, particularly when I was in Engineering Excellence, I observed a resistance to advice if it was gleaned from experience within Microsoft itself. There was a team in Microsoft Research that examined software engineering, generally using Microsoft as its population of interest, but the results, although considered to be thought-provoking tidbits, rarely had any uptake back in the product groups. It was a variation of the Gell-Mann Amnesia effect, where people realize that news stories about their areas of expertise are simplistic or inaccurate, but completely trust news stories about topics they know nothing about. If you told members of one Microsoft team about the engineering experience of another team, they would immediately be able to identify—because of their knowledge of the internals of Microsoft—the ways in which that other team was different from their team, and therefore dismiss the guidance as not relevant. Meanwhile, they would happily slurp up guidance on Scrum, even if it was completely inapplicable to their team, because they weren't aware of the details of the environment in which it had been successful.

Set a Goal of Eventual Certification and Licensing

McConnell's 1999 book *After the Gold Rush: Creating a True Profession of Software Engineering* addresses a theme similar to that of this book: How do we fix software engineering? He references an earlier iteration of SWEBOK as a guide for what universities should teach students, and proposes the certification (voluntary) or licensing (mandatory) of software engineers.[42]

I concur with McConnell's ultimate goal: software engineering should be taught from an agreed-on body of knowledge, which could then lead to the certification and licensing of software engineers based on that body of knowledge. This would benefit everybody—colleges would know what to teach, companies would know what to interview for, and software developers would know the gaps in their skills, and of course, in the end, they would do a better job and we would have more reliable software.

Unfortunately, SWEBOK is still not yet ready for prime time; it includes curriculum guidelines, but I agree with the SEMAT book's dismissive summary of those: "guidelines at a very high level, leaving (too) much to be defined by individual universities and professors."[43] In 2013, the Texas Board of Professional Engineers, as part of the licensing of software engineers, adopted the Principles and Practice of Engineering exam, which exists for a wide variety of disciplines, in the area of software engineering.[44] The exam, which covers a broad but shallow SWEBOK-ish range of knowledge, is being viewed with a "wait-and-see" attitude by other jurisdictions.[45]

Even with a well-defined body of knowledge, companies' attitude toward certification and licensing would likely still be indifferent at best, and actively oppositional at worst, because they view them as having no tangible benefits. Companies care about employees knowing the specific languages and tools they need for their job, but those vary widely across the industry. In addition, they are concerned that licensing would open up the issue of liability for bugs.

Parnas provided a version of a standard line (attributing it to a friend): "We live in countries where you need a license to cut hair, but you don't need anything to write code for safety critical software or other mission critical software."[46] McConnell helpfully supplied a list of over thirty professions that require licensing in the state of California, including custom upholsterer and mule jockey.[47] It's an easy punch to land, but it's still true.

Certification and licensing should happen, but not right away; what is important is that industry and academia agree that they are a goal to strive for. Maybe it's not required for everybody, and maybe you need a master's degree, but we should set it as a long-term goal. In 1999, McConnell wrote, "We need to continue working on several fronts—instituting widespread undergraduate education, licensing professional software engineers, establishing software engineering certification programs, and thoroughly diffusing best practices into the industry."[48] Twenty years later, there has been almost zero progress in those areas.

In the meantime, what do we have? What we have is the situation I found myself in, twenty-plus years into my career as a software developer: I realized that there was a curious void in my professional life because there was no craftsperson-apprentice relationship that I could develop with recent college graduates. I could mentor people on how to navigate the waters of corporate life, but that was generic advice that they could get from anybody.

Like others, my guidance was vague: "Well, in this one case I remember this sort of thing worked OK, so why not try that?"

I saw the author Michael Lewis give a talk at Microsoft about his book *The Big Short*, which covered the financial meltdown of 2008. Lewis's first book, *Liar's Poker*, chronicled his time working on Wall Street, the headquarters of the US financial services industry. He mentioned that over the years, many people told him they went to work on Wall Street after reading *Liar's Poker*. He found this curious because his intent had been to demonstrate that Wall Street was a silly place to work, but somehow it came across to college students as an interesting place to work. He realized that an author doesn't have control over readers, but he did feel a little funny because he felt that Wall Street was a huge waste of talent.

Is the software industry a huge waste of talent? I am sure that most Wall Street workers feel that they provide a valuable service, notwithstanding Lewis's opinion. And I certainly don't feel that my career has been a waste. I've worked on software products that are extremely useful to many people. A few employees have told me that reading one of my earlier books inspired them to want to work for Microsoft, which made me proud, not guilty. Still, I have felt the effect of a problem that a Microsoft executive once lamented: "We hire a lot of smart people at Microsoft, but they tend to cancel each other out." Expending time and energy, when I was working on product teams at Microsoft, convincing peers of the benefits of a certain approach to the software development process (or worse, failing to convince them because their unexamined beliefs were as entrenched as mine), when the positives or negatives of such a process should have been worked out long ago, was extremely wasteful, not to mention frustrating. I think of the first line of Allen Ginsberg's poem "Howl," "I saw the best minds of my generation destroyed by madness," and I feel some of his frustration.[49]

Steve Lohr, in his history of software titled *Go To*, stated the proposition that "Americans typically brought an engineering mentality to the task of designing programming languages—compromises were made to solve the computing problem at hand. The Europeans, by contrast, often took a more theoretical academic approach to language design."[50] This claim, which he admits is a "broad generalization," is not borne out by the facts. You can look back at the history of programming languages and find similar ideas being propagated on both sides of the Atlantic. Simula and Pascal came from Europe, and Smalltalk and Algol came from the United States. The two early salespeople of object-oriented design, Meyer and Cox, were French and American, respectively. Dijkstra was European, and Knuth is American. A Dane working in the United States designed C++, the ultimate compromise. If the availability of peanut butter has an effect on language design, I haven't observed it.

What is undoubtedly true, however, is that as the needs of industry outstripped the ability of college professors to keep up, the nexus of software engineering moved firmly to the United States, because that's where almost all the largest software companies are headquartered.

Is the current state of software a reflection of US values that favor individuality over community, breaking the rules over following them, MacGyver over Hercule Poirot? It's not that US companies institutionally ignore quality; the hardware companies that initially fostered an engineering-focused approach to software, when they were the places where software was being written back in the 1960s, were also primarily American. Kurt Andersen, in his recent book *Fantasyland*, discusses how starting in the 1960s Americans began to drift away from rationalism, as individuals increasingly felt that *"your beliefs* about anything are equally as true as anyone else's ... *What I believe is true because I want and feel it to be true."*[51] This would make software the quintessential American industry, and certainly there is something about the speed with which software has arrived on the scene, coupled with the amount of money being made today, that seems to uniquely bind software to foundational US legends—the self-made person, the genius in the garage, the myth of the West. It may be impossible to convince US software leaders that they aren't experiencing manifest destiny replayed in the twenty-first century, and that the software industry should be anything other than exactly what it is now. If you use the number of millionaires as your metric for determining the success of an industry, software engineering appears to be doing great.

But that doesn't mean we don't need to change it. And the change doesn't have to originate in the United States. There are plenty of technologies, from cars to skyscrapers to solar panels, which initially flowered in the United States but whose center of innovation later shifted elsewhere. Currently, the United States is the predominant source of wisdom, as it were, on software development, but there is no innate reason that it has to stay that way, particularly if other countries are more willing to invest in fundamental engineering research.

Mills fretted about this back in 1980, referring to the missile gap that existed with the Soviet Union in the early days of the Cold War: "The inertia of several hundred thousand undisciplined programmers in the United States is a real reason for concern. ... Unless we address this problem with exceptional measures, we are on the way to a 'software gap' much more serious and persistent than the famous 'missile gap' which helped fuel the very growth of our electronics industry."[52] So far this has not manifested itself: the mass of undisciplined programmers has grown, but they are not uniquely concentrated in the United States.

Meanwhile, noted skeptic Edward Yourdon (whose early warnings about the impending Y2K catastrophe, depending on whom you ask, either vastly overstated the problem or gave the world time to avoid it) published a book in 1992 titled *Decline and Fall of the American Programmer*, in which he blithely predicted, "The American programmer is about to share the fate of the dodo bird. By the end of this decade, I foresee massive unemployment among the ranks of American programmers, systems analysts, and software engineers." His premise was that "American software is developed at a higher cost, less productively, and with less quality than that of several other countries." Therefore, "if your slovenly bunch of software engineers doesn't want to play at the world-class level of performance, trade them in for a new bunch from Ireland or Singapore."[53]

This has not come to pass. US code is currently no worse than other code. Yourdon himself, a mere four years later, published the book *Rise and Resurrection of the American Programmer*, in which he walked back some of his earlier warnings, inspired by, among other things, the growth of the Internet and success of Microsoft. Referring to the strawperson COBOL programmer from his first book, Yourdon wrote, "*That* American programmer is indeed dead, or at least in grave peril. But there's a new generation of American programmers, doing exciting new things."[54] But of course that doesn't mean it can't happen in the future. Personally I don't care; I want software to become a real engineering discipline, but this is unrelated to whether the center remains in the United States.

Coming back to the original question I asked in the introduction, Is software development really hard, or are software developers not that good at it? Certainly there are parts of software that are hard, but software developers seem to do everything in their power to make even the easy parts harder by wasting an inordinate amount of time on reinvention and inefficient approaches. A lot of mistakes are in fundamental areas that should be understood by now, and so the software industry needs to push to figure them out and then teach people about them, so we can devote our energy to the parts that really are hard.

In the 1982 coming-of-age movie *Fast Times at Ridgemont High*, the character of Jeff Spicoli is visited on the evening of his high school graduation dance by his history teacher, the inimitable Mr. Hand, and forced to atone for his previously lackluster commitment to scholarship by demonstrating his knowledge of the American Revolution. At the end, Spicoli produces this summary: "What Jefferson was saying was, hey, we left this England place because it was bogus. So if we don't get some cool rules ourselves, pronto, we'll just be bogus too."[55]

This is what software engineering is facing today. We left behind the other jobs we could have taken because they didn't appeal to us, and headed to the land of software where we could be clever and creative. We had fun for a while, but now the whole world is depending on us, and if we don't get some cool rules ourselves, pronto, ... well, you know.

Notes

Introduction

1. Mary Shaw, "Prospects for an Engineering Discipline of Software," *IEEE Software* 7, no. 6 (November 1990): 16, 18.

2. Ibid., 21.

3. Ibid., 24.

4. Greg Wilson, "Two Solitudes" (keynote address, SPLASH 2013, Indianapolis, IN, October 30, 2013), accessed December 18, 2017, https://www.slideshare.net/gvwilson/splash-2013.

5. Greg Wilson and Jorge Aranda, "Two Solitudes Illustrated," December 6, 2012, accessed December 18, 2017, http://third-bit.com/2012/12/06/two-solitudes-illustrated.html.

Chapter 1

1. It's a Canadian catalog, so the actual prices are in Canadian dollars; I've approximated them with an exchange rate of US80¢ for one Canadian dollar, which is roughly right for early 1982. The actual prices, in Canadian dollars, are $999, $1,399, and $2,899 for the computers, $6,295 for the hard disk, and $1,199 for one hard drive with TRSDOS and Disk BASIC.

2. Adam Bryant, "Wang Files for Bankruptcy; 5,000 Jobs to Be Cut," *New York Times*, August 19, 1992, accessed December 19, 2017, http://www.nytimes.com/1992/08/19/business/wang-files-for-bankruptcy-5000-jobs-to-be-cut.html.

3. Hewlett-Packard, *HP-41C/41CV Owner's Handbook and Programming Guide* (Corvallis, OR: Hewlett-Packard, 1982), 7. We did not own this calculator, but I happened to find an HP-41CV, along with the manual, abandoned in a Microsoft office that I was moving into in 2010.

4. Intel Corporation, *Intel 80386 Programmer's Reference Manual 1986* (Santa Clara, CA: Intel, 1987), 244, 261.

5. That instruction takes up two bytes, with the first one,

`0  0  0  0  0  0  0  1`

telling the processor that this is an ADD instruction, and the second one,

`1  1  0  0  0  0  0  1`

telling the processor that it should add EAX into ECX—and all this is encoded in those bits according to the rules of the processor. For the record, the initial 1 1 tells the processor that this is a register-to-register add, the next 0 0 0 identifies the source as EAX, and the 0 0 1 identifies the target as ECX.

6. Also, somewhat confusingly, the main memory on the calculator was referred to as registers in the HP manuals.

7. $299 in Canadian dollars.

8. WATFIV was developed at the University of Waterloo in the late 1960s; it succeeded a version known as WATFOR, for "Waterloo Fortran," and the name was chosen to be "the one after WATFOR," although it is also a short form of "Waterloo Fortran IV." Fortran IV, in turn, was the fourth version of Fortran. Paul Cress, Paul Dirksen, and J. Wesley Graham, *FORTRAN IV with WATFOR and WATFIV* (Englewood Cliffs, NJ: Prentice-Hall, 1970), v.

9. Ibid., 62.

10. "WATFIV User's Guide," accessed December 19, 2017, http://www.jaymoseley .com/hercules/downloads/pdf/WATFIV_User_Guide.pdf.

11. You could buy a model that only displayed text, but we had the more advanced one.

12. IBM, *BASIC 1.10* (Boca Raton, FL: IBM, 1982), 4-212. L8 makes each note an eighth note, a minus sign after a note makes it flat, a 4 after a note makes it a quarter note, P8 pauses for an eighth note, and spaces are ignored. IBM, *BASIC 1.00* (Boca Raton, FL: IBM, 1981), 4-180–4-181.

13. IBM, *BASIC 1.00*, 4-71–4-72. Mx,y moves to coordinates x,y; R/D/L/U draws to the right/down/left/up; and E and F draw diagonally up and right and diagonally down and right, respectively.

14. Matthew Reed, "Level I Basic," TRS-80.org, accessed December 19, 2017, http:// www.trs-80.org/level-1-basic/.

15. Steve Teglovic Jr. and Kenneth D. Douglas, *Structured BASIC: A Modular Approach for the PDP-11 and VAX-11* (Homewood, IL: Richard D. Irwin, 1983), 160.

16. BASIC did have the ability to define "functions," which took some number of parameters and returned a single value, but they were restricted to a single line of code, and were meant for simple mathematical or string manipulation functions, not general parameterized subroutines. For example, you could write DEF FNAREA(R) = 3.14 * (R ^ 2) (with ^ meaning "raised to the power of") and thenceforth use FNAREA(X) for any variable X.

17. IBM PC BASIC could load code from another program via the obscure CHAIN MERGE API and it suffered from exactly this danger.

18. The source code for all the IBM PC DOS BASIC samples is available on Leon Peyre's website; search for "DOS 1.1 Samples." He also has some of the games from the *Basic Computer Games* books. This reveals that at its core, the city skyline game (named ART.BAS) is in fact eleven lines, although three of them are for playing sounds. Leon Peyre, "Back to BASICs: A Page about GWBASIC Games and Other Programs," accessed December 20, 2017, http://peyre.x10.mx/GWBASIC/.

19. Glenn Dardick, Facebook messenger exchange with the author, June 13, 2017.

20. Bill Gates, remarks at Tech Ed 2001 conference, June 19, 2001, accessed December 20, 2017, https://web.archive.org/web/20070704104845/http://www.microsoft .com/presspass/exec/billg/speeches/2001/06-19teched.aspx.

21. Technically it was 16 bit, but you know what I mean. The Wikipedia page has links to the source code as well as a video of the game in action and playable versions for modern computers. "DONKEY.BAS," accessed December 20, 2017, https:// en.wikipedia.org/wiki/DONKEY.BAS.

22. "donkey.bas," accessed December 20, 2017, https://github.com/coding-horror/ donkey.bas/blob/master/donkey.bas.

23. Ibid.

24. Ibid.

25. Ibid.

26. Bill Crider, ed., *BASIC Program Conversions* (Tucson: HPBooks, 1984).

27. IBM, *BASIC 1.00*, D-2.

28. David H. Ahl, ed., *Basic Computer Games: Microcomputer Edition* (New York: Workman Publishing, 1978), xii.

29. Ibid., 90.

30. Ibid., 157–163; David H. Ahl, ed., *More Basic Computer Games* (New York: Workman Publishing, 1979), 143–149.

31. Vincent Erickson, e-mail exchange with the author, July 8, 2017.

32. Ahl, *Basic Computer Games*, 90.

33. John G. Kemeny and Thomas E. Kurtz, *Back to Basic: The History, Corruption, and Future of the Language* (Reading, MA: Addison-Wesley, 1985), 43–53.

34. Ibid., 14, 63.

35. Edsger W. Dijkstra, "How Do We Tell Truths That Might Hurt?" (June 18, 1975), in *Selected Writing on Computing: A Personal Perspective* (New York: Springer Verlag, 1982), 129–131.

Chapter 2

1. Vaughan Pratt, interview with the author, February 7, 2017; Donald Knuth, interview with the author, May 18, 2017. Both mentioned this about Princeton.

2. Edsger W. Dijkstra, "How Do We Tell Truths That Might Hurt?" (June 18, 1975), in *Selected Writing on Computing: A Personal Perspective* (New York: Springer Verlag, 1982), 130.

3. Peter Grogono, *Programming in Pascal*, 2nd ed. (Reading, MA: Addison-Wesley, 1984).

4. Harlan D. Mills, "Reading Code as a Managerial Activity," in *Software Productivity* (New York: Dorset House, 1988), 181–182.

5. Donald E. Knuth, "Literate Programming," *The Computer Journal* 27, no. 2 (January 1984): 97.

6. Gerard M. Weinberg, foreword to *Structured Programming in APL*, by Dennis P. Geller and Daniel P. Freedman (Cambridge, MA: Winthrop Publishers, 1976), xi.

7. Robert T. Grauer and Marshal A. Crawford, *Structured COBOL: A Pragmatic Approach* (Englewood Cliffs, NJ: Prentice-Hall, 1981), 95.

8. Steve Teglovic Jr. and Kenneth D. Douglas, *Structured Basic: A Modular Approach for the PDP-11 and VAX-11* (Homewood, IL: Richard D. Irwin, 1983), 114.

9. Ibid., 119.

10. Dennis P. Geller and Daniel P. Freedman, *Structured Programming in APL* (Cambridge, MA: Winthrop Publishers, 1976), 282.

11. Ibid., 283.

12. J. N. P. Hume and R. C. Holt, *Structured Programming Using PL/1 and SP/k* (Reston, VA: Reston Publishing, 1975), 2–3. The name of the language is usually written as PL/I, using a capital letter *I* to indicate "one," but this book uses PL/1 with the digit *1*.

13. Ibid., 76.

14. Grauer and Crawford, *Structured COBOL*, 97.

15. Edsger W. Dijkstra, "From 'Goto Considered Harmful' to Structured Programming," in *Software Pioneers: Contributions to Software Engineering*, ed. Manfred Broy and Ernst Denert (Berlin: Springer Verlag, 2002), 346.

16. Edsger W. Dijkstra, "Letters to the Editor: Go To Statement Considered Harmful," *Communications of the ACM* 11, no. 3 (March 1968): 147–148.

17. Ibid.

18. Ibid.

19. Harlan D. Mills, "Mathematical Foundations for Structured Programming," in *Software Productivity* (New York: Dorset House, 1988), 120.

20. Corrado Böhm and Giuseppe Jacopini, "Flow Diagrams, Turing Machines, and Languages with Only Two Formation Rules," *Communications of the ACM* 9, no. 5 (May 1966): 366–371.

21. Grogono, *Programming in Pascal*, 300–301.

22. BASIC line numbers weren't required to increment by 10, but they traditionally did; line 690 here represents "the line before line 700."

23. John G. Kemeny and Thomas E. Kurtz, *Back to Basic: The History, Corruption, and Future of the Language* (Reading, MA: Addison-Wesley, 1985), 82.

24. Henry F. Ledgard and Louis J. Chmura, *Fortran with Style: Programming Proverbs* (Rochelle Park, NJ: Hayden, 1978), 20.

25. Harlan D. Mills, "The Case against GO TO Statements in PL/I," in *Software Productivity* (New York: Dorset House, 1988), 27.

26. Ibid., 28.

27. Ibid., 27.

28. Vladimir Zwass, *Programming in FORTRAN: Structured Programming with FORTRAN IV and FORTRAN 77* (New York: Barnes and Noble, 1981), 13–14.

29. Ibid., 132–137.

30. Ibid., 111.

31. Grauer and Crawford, *Structured COBOL*, 114.

32. Ibid., 201, 191.

33. Donald E. Knuth, "Structured Programming with Go To Statements," *Computing Surveys* 6, no. 4 (December 1974): 269; Dijkstra, "Letters to the Editor: Go To Statement Considered Harmful," 147, 148.

34. Raymond M. Smullyan, *What Is the Name of This Book?: The Riddle of Dracula and Other Logical Puzzles* (New York: Simon and Schuster, 1978), 206.

35. Jane Margolis and Allan Fisher, *Unlocking the Clubhouse: Women in Computing* (Cambridge, MA: MIT Press, 2002), 79, 81, 101.

36. Ibid., 4, 16–17.

37. Ibid., 39. Here they are referencing two books: Carol Gilligan, *In a Different Voice: Psychological Theory and Women's Development* (Cambridge, MA: Harvard University Press, 1982); Lyn Mikel Brown and Carol Gilligan, *Meeting at the Crossroads: Women's Psychology and Girls' Development* (Cambridge, MA: Harvard University Press, 1992).

38. Ibid., 33.

39. This number may be slightly inaccurate since you could choose a combined electrical engineering and computer science major, so it is hard to tell who was focused on computer science.

40. Abraham Flexner, "Medical Education in the United States and Canada: A Report to the Carnegie Foundation for the Advancement of Teaching," *Carnegie Foundation Bulletin* 4 (1910): 3. Originally published by Merrymount Press, Boston, reproduced by photolithography by W. M. Fell, 1960, accessed December 22, 2017, http://archive.carnegiefoundation.org/pdfs/elibrary/Carnegie_Flexner_Report.pdf.

41. David Alan Grier, "The Migration to the Middle," *IEEE Computer* 44, no. 1 (January 2011): 13–14.

42. Harlan D. Mills, "Software Development," in *Software Productivity* (New York: Dorset House, 1988), 231–232.

43. Gerald M. Weinberg, *The Psychology of Computer Programming*, silver anniversary ed. (New York: Dorset House, 1998), 150.

Chapter 3

1. If you want to try this at home, Microsoft provides free tools. First, search for and download Visual Studio Community, and then run it (it takes several minutes). Then, in the version I used in early 2018, you have to select .NET Desktop Development from the install window. When the install is finished, launch Visual Studio (no need to sign in if prompted, and keep the default settings). From the File menu, choose New ... and then Project. Create a Console App using the .NET Framework in Visual C# (you can use the default name, which was ConsoleApp1 on my version). Next, on the right side, in the Solution Explorer window, right-click on Reference, select Add Reference ..., and choose Systems.Windows.Forms (you have to click in the box to the left to really select it), and then OK. Near the top of the code in the

main `Program.cs` window add a new line that reads `using  System.Windows.` `Forms;` after the other lines that start with "using." Paste the code sample from this book between the most-indented { and } lines (the ones that follow the line `static` `void Main(string[] args)`). Hit F5 to compile and run the program (or choose Debug and then Start Debugging). If you get an error, make sure you have typed everything correctly, including all punctuation symbols as well as preserving upper- and lowercase—and you are now experiencing the thrill of debugging! Don't forget to hit "OK" in the message box to dismiss it, so the program finishes running or else you won't be able to make further changes to the code.

2. Frederick P. Brooks Jr., "The Tar Pit," in *The Mythical Man-Month: Essays on Software Engineering*, anniversary ed. (Boston: Addison-Wesley, 1995), 7–8.

3. Fred Moody, *I Sing the Body Electronic: A Year with Microsoft on the Multimedia Frontier* (New York: Viking, 1995), 125–126.

4. Barry Schwartz, *The Paradox of Choice: Why More Is Less*, rev. ed. (New York: HarperCollins, 2016), 103.

5. Robert L. Grady, *Successful Software Process Improvement* (Upper Saddle River, NJ: Prentice-Hall, 1997), 8.

6. John Shore, "Myths of Correctness," in *The Sachertorte Algorithm and Other Antidotes to Computer Anxiety* (New York: Viking, 1985), 175.

7. André van der Hoek and Marian Petre, "Postscript," in *Software Designers in Action: A Human-Centric Look at Design Work* (Boca Raton, FL: CRC Press, 2014), 403.

8. This was a minor plot point on an episode of the television show *Silicon Valley*, when a programmer decides he can't date someone who uses spaces instead of tabs.

9. Harlan D. Mills, "In Retrospect," in *Software Productivity* (New York: Dorset House, 1988), 3.

10. Frederick P. Brooks Jr., "The Whole and the Parts," in *The Mythical Man-Month: Essays on Software Engineering*, anniversary ed. (Boston: Addison-Wesley, 1995), 142.

11. It's not clear why this style is associated with the language Pascal; it may simply have been the first programming language that inspired programmers to write in mixed case at all. Certainly Niklaus Wirth, the inventor of the language, did not use that style in his original paper on the language. He demonstrates a procedure named Bisect, but also one named readinteger. Niklaus Wirth, "The Programming Language Pascal," *Acta Informatica* 1, no. 1 (1971): 35–63.

12. Joel Spolsky, "Making Wrong Code Look Wrong," *Joel on Software* (blog), May 11, 2005, accessed December 29, 2017, https://www.joelonsoftware.com/2005/05/11/making-wrong-code-look-wrong/.

13. Larry Osterman, "Hugarian Notation—It's My Turn Now :)," *Larry Osterman's WebLog* (blog), June 22, 2004, accessed December 27, 2017, https://blogs.msdn .microsoft.com/larryosterman/2004/06/22/hugarian-notation-its-my-turn-now/; Scott Ludwig, comment on Osterman, "Hugarian Notation," accessed December 27, 2017, https://blogs.msdn.microsoft.com/larryosterman/2004/06/22/hugarian -notation-its-my-turn-now/#comment-7981.

14. G. Pascal Zachary, *Showstopper: The Breakneck Race to Create Windows NT and the Next Generation at Microsoft* (New York: Free Press, 1994), 56.

15. Ibid.

16. Microsoft, "Compiler Error CS0845," July 20, 2015, accessed December 27, 2017, https://docs.microsoft.com/en-us/dotnet/csharp/language-reference/compiler -messages/cs0845.

17. Please note a couple of things. First, if you are typing this code at home and want it to compile, you have to add a line

```
using System.Globalization;
```

at the top of your source code file, and you also have to use `CultureInfo.Invari-antCulture` instead of plain `InvariantCulture` (or you could call the API `ToUp-perInvariant()`, which does the exact same thing). Second, it is a best practice, when doing comparisons of this sort, to avoid doing uppercasing yourself, so instead of first calling `ToUpper()` and then `Contains()`, you should call an API directly on `ErrorMessage`:

```
if (ErrorMessage.IndexOf("javascript",
    StringComparison.InvariantCultureIgnoreCase) != -1)
```

The reason you have to call `IndexOf()` instead of `Contains()` is that `Contains()` does not have an overload that lets you specify `InvariantCultureIgnoreCase`. Why doesn't it have this? I don't know; once again you are at the mercy of the API designer.

18. Brooks, "Tar Pit," 4, 6.

19. Ibid., 5–6.

20. Carl Landwehr, Jochen Ludewig, Robert Meersman, David Parnas, Peretz Shoval, Yair Wand, David Weiss, and Elaine Weyuker, "Systems Software Engineering Programmes: A Capability Approach," *Journal of Systems and Software* 125 (2017): 354–364. This article relates Brooks's explanation to the gap in current software engineering education.

21. Brooks, "Tar Pit," 9.

Chapter 4

1. The proper capitalization of UNIX inspires some debate. The name is not an acronym, but is often written in uppercase. According to the Jargon File website, the original article about it (Dennis M. Ritchie and Ken Thompson, "The UNIX Time-Sharing System," *Communications of the ACM* 17, no. 7 [July 1974]: 365–375) only used capital letters—actually small ones—because the authors had access to a new typesetting system and were "intoxicated by being able to produce small caps." Ritchie later tried unsuccessfully to get the official spelling changed to "Unix." In honor of this, the Jargon File itself uses "Unix." Jargon File entry for "Unix," accessed December 29, 2017, http://catb.org/jargon/html/U/Unix.html.

2. Jon Bentley, *Programming Pearls* (Reading, MA: Addison-Wesley, 1986), 151.

3. Jon Bentley, "Squeezing Space," in *Programming Pearls* (Reading, MA: Addison-Wesley, 1986), 93.

4. Jon Bentley, *Writing Efficient Programs* (Englewood Cliffs, NJ: Prentice-Hall, 1982), xii.

5. Dennis M. Ritchie, "The Development of the C Language," accessed December 29, 2017, https://www.bell-labs.com/usr/dmr/www/chist.html.

6. Although for portability reasons, it is intentionally not tied to any specific assembly language; C compilers can be written for any processor.

7. Certain older computers, certainly at the time that C was invented, had different numbers of bits in a byte (for example, the original Kernighan and Ritchie C book talks about a Honeywell computer that has 9 bits in a byte), but today 8 bits per byte is the standard. Brian W. Kernighan and Dennis M. Ritchie, *The C Programming Language* (Englewood Cliffs, NJ: Prentice-Hall, 1978), 34.

8. For details on why the range goes from −128 to 127 instead of −127 to 127, see Wikipedia, "Two's Complement," accessed December 29, 2017, https://en.wikipedia.org/wiki/Two%27s_complement. Of course, −128 to 128 wouldn't fit in 8 bits, since that is 257 distinct numbers and there are only 256 unique 8-bit numbers.

9. Yes, it's actually slightly more than the 32-bit number squared. The 64-bit range is 0 to 18,446,744,073,709,551,615 (or -9,223,372,036,854,775,808 to 9,223,372,036,854,775,807). If you are a fan of US college basketball, you may have heard the number 9 quintillion before, since it is stated as the chance of perfectly guessing the bracket for the National Collegiate Athletic Association's Division I basketball championship, which is held each spring. It's not a coincidence that this is also the largest signed 64-bit number. The tournament is a 64-team single-elimination one; to knock out sixty-three teams and leave one champion, you need sixty-three games, and since each of the games has two possible outcomes, the chance of getting every one right is 2^{63}, which is that 9 quintillion number. In recent

years, though, the tournament has expanded to sixty-eight teams, so in fact the chance of guessing everything right is 2^{67}, which is sixteen times larger.

10. Computers, certainly all modern ones as well as many of the ones that existed back in the 1970s, also supported *floating point* numbers, which stored numbers in what we think of as scientific notation: a number and decimal fraction followed by a power of ten that they were multiplied by. This led to a much larger range of numbers, but also removed the precision of using integers; there was a bit of a kerfuffle in the early days of the IBM PC when somebody discovered that (1/3) * 3 did not equal exactly 1 when using floating point numbers in BASIC.

11. There was one exception to the "integer is the natural processor bitness" rule: although the early personal computers from Apple, Radio Shack, and Commodore were all 8-bit computers, they all supported 16-bit integers; restricting an integer to the range –128 to 127 was too limiting, and too likely to cause a program termination due to overflow, to be useful, so BASIC did the extra work to handle 16-bit integers, ranging from –32,768 to 32,767. That isn't a huge range either—certainly a spreadsheet, say, could store a number bigger than that—but it worked well enough for hobby-level programming. In addition, some versions of BASIC used floating point numbers rather than integers by default, with the expanded range but lack of precision, as discussed above. Bill Crider, ed., *BASIC Program Conversions* (Tucson: HPBooks, 1984), 21–22.

12. IBM, *Basic 1.00* (Boca Raton, FL: IBM, 1982), A-4.

13. As mentioned in a previous note, the default int did not have a strictly defined bitness. It was supposed to "reflect the 'natural' size for a particular machine" (Kernighan and Ritchie, *C Programming Language*, 34). Furthermore, the char type, which was the 8-bit number on most machines, was originally only defined as "capable of holding one character in the local character set" (ibid.). Meanwhile, short and long, the 16- and 32-bit numbers on most machines (with long being introduced a bit later in the language's history, with the original PDP-11 machine that it was developed on being only a 16-bit machine, and the VAX, the successor to the PDP that supported 32 bit, not coming out until 1977), were left up to the compiler writer to interpret, with the caveat that short should not be bigger than long. In practice, char, short, and long meant 8, 16, and 32 bits, and when the C language was standardized by ANSI in 1990, those definitions became official.

14. If you are worried, you can write code that checks for overflow before it happens (you can't directly implement the "after the fact" check in C, because it doesn't expose the "did the last operation overflow" information from the processor, so the precheck code has to do extra work before performing the operation, which winds up being slower). In practice, programmers size their variables as large as they think they need to be, and thereafter ignore the possibility of overflow.

15. Nancy G. Leveson and Clark S. Turner, "An Investigation of the Therac-25 Accidents," *IEEE Computer* 26, no. 7 (July 1993): 34; James Gleick, "A Bug and a Crash," accessed December 30, 2017, https://around.com/ariane.html.

16. Kernighan and Ritchie, *C Programming Language*, 89.

17. Ritchie, "Development of the C Language."

18. Although multiplying by a number that is a power of two, such as two, four, or eight—which are the byte sizes of the standard C numerical types that arrays are often filled with—is faster than multiplying by an arbitrary number.

19. IBM, *Basic 1.00*, A-5.

20. Presumably, the reason that IBM PC BASIC strings were limited to precisely 255 characters is because internally, it was using an 8-bit unsigned number to store the length.

21. Operating systems tended to round up memory allocation requests to a multiple of a power of two, such as four or sixteen, so you might get lucky and have a few extra bytes to play with, but this would depend on the exact length of the string in that particular run of your program.

22. The problem was, If you consider two 5-letter words to be "equivalent" if they differ only in one letter, and a word is put into a bucket as long as it is "equivalent" to another word in the bucket, how many distinct buckets are there in this list of 5-letter words?—the list of words, of course, not being known before we submitted our programs.

23. Ted Eisenberg, David Gries, Juris Hartmanis, Don Holcomb, M. Stuart Lynn, and Thomas Santoro, "The Cornell Commission: On Morris and the Worm," *Communications of the ACM* 32, no. 6 (June 1989): 706–709.

24. Donn Seeley, "A Tour of the Worm," February 1989, accessed December 30, 2017, https://collections.lib.utah.edu/details?id=702918; Bob Page, "A Report on the Internet Worm," November 7, 1988, accessed December 30, 2017, http://www.ee.ryerson.ca:8080/~elf/hack/iworm.html.

25. Joyce Reynolds, "The Helminthiasis of the Internet," *Computer Networks and ISDN Systems* 22, no. 4 (October 1991): 347–361. The title is a nerd joke; *helminthiasis* is a medical term for "infestation with parasitic worms."

26. David Zimmerman, "The Finger User Information Protocol," Internet Network Working Group Request for Comments: 1288, December 1991, accessed December 30, 2017, https://tools.ietf.org/html/rfc1288. This provides an approachable peek into what formal Internet protocol specifications look like. It also features, in section 4 at the end, an example of an actual finger command, which immortalizes the author's coworkers circa 1991. When finger is asked to report back all users on a system, it shows terminal location, office location, office phone number, job name,

and idle time (the number of minutes since either the last typed input or job activity). When asked about a specific user, it returns office location, office phone number, home phone number (!), log-in status (not logged in, logout time, etc.), and the .plan and .project file contents.

27. Victoria Neufeldt, ed., *Webster's New World Dictionary of American English*, 3rd college ed. (New York: Prentice Hall, 1991), 347.

28. You can allocate memory off the heap until the operating system realizes that it is all used up (or more precisely, that it can't find a consecutive run of unallocated bytes that is large enough to satisfy your request; a previous 1-byte allocation in the middle of your 1 megabyte of memory would make it impossible to allocate more than half a megabyte of memory at once). At this point, it will return a value of 0 as the result of the allocation, which is an invalid pointer. Note also that the operating system does keep track of the size of each heap allocation, so it can free it properly, but this information is not available to C code.

29. In addition to storing parameters, return addresses, and local variables, the stack is used to save the values of a few of the processor registers at the beginning of each function.

30. This is a simplification; on certain processors, some parameters are passed in registers rather than on the stack, and the compiler may choose to optimize by storing a local variable in a register versus on the stack. A large buffer, however, would always go on the stack.

31. I don't have the source code to the finger daemon that was running at the time; I am relying on investigations into the virus published at the time for the detail that gets() was the vector for the buffer overflow. It is not clear why the code would be written that way; it would mean that the finger daemon took the data in the packet and then called another program to deal with it, and that other program called gets(). It seems a bit overdesigned for something as simple as finger. Still it's possible, especially if the finger daemon was actually a subset of a more generic daemon configured to receive more generic packets and then feed them to a specific program to respond.

32. fgets() has another difference in that it may copy an extra character—the "end of line" character—into the buffer, so the daemon may need to write a little bit of code to handle that as well. This is another unexplainable difference between gets() and fgets(). See, among others, Eugene Spafford, "The Internet Worm Program: An Analysis," accessed January 1, 2018, http://spaf.cerias.purdue.edu/tech -reps/823.pdf.

33. The finger daemon can optionally support chaining requests together, so you can ask one machine to forward your request on to a third machine, to an arbitrary length; this is much less useful now, when almost every machine is directly accessible from every other machine, but technically it means 64 bytes might be too short.

34. Trend Micro, "TrendLabs SECURITY INTELLIGENCE Blog (June 2, 2017), accessed January 2, 2018, http://blog.trendmicro.com/trendlabs-security-intel ligence/ms17-010-eternalblue/.

35. J. R. R. Tolkien, *The Return of the King* (Boston: Houghton Mifflin, 2002), 1074. She was referring to the unpredictable yet inevitable mortality of humankind, although the quote also applies to the unpredictable yet inevitable exploitability of buffer handling code in C.

36. Ralf Burger, *Computer Viruses: A High-Tech Disease* (Grand Rapids, MI: Abacus, 1988).

37. Ibid., 187–189.

38. Ibid. Even a quick scan of the code reveals that something is up. After an initial comment, there are twenty-two lines of code to print out the tree itself, each calling the charmingly named SAY API, which is REXX's way of printing out text, with each line of the asterisk tree visible as the parameter, then a comment saying, "Browsing this file is no fun at all," which appears to be an attempt to get people to look no further, and then sixty more lines, which clearly have nothing to do with printing the tree, since that has already been done.

39. Spafford, "Internet Worm Program," 23.

40. Eisenberg et al., "Cornell Commission," 709.

41. Niklaus Wirth, "A Brief History of Software Engineering," *IEEE Annals of the History of Computing* 30, no. 3 (July–September 2008): 34.

42. Ibid.

43. Seeley, "Tour of the Worm," 8.

Chapter 5

1. Matti Tedre, *The Science of Computing: Shaping a Discipline* (Boca Raton, FL: CRC Press, 2015), 120.

2. Peter Naur, Brian Randell, and J. N. Buxton, ed., *Software Engineering Concepts and Techniques: Proceedings of the NATO Conferences* (New York: Petrocelli/Charter, 1976).

3. Tedre, *Science of Computing*, 122. He is referencing Naur, Randell, and Buxton, *Software Engineering Concepts and Techniques*, 145.

4. Tedre, *Science of Computing*, 124. He is referencing Thomas Haigh, "Dijkstra's Crisis: The End of Algol and the Beginning of Software Engineering: 1968–1972" (paper presented at the History of Software, European Styles conference, Lorentz Center, University of Leiden, Netherlands, September 17, 2010).

5. Alan Cooper, *The Inmates Are Running the Asylum: Why High Tech Products Drive Us Crazy and How to Restore the Sanity* (Indianapolis: Sams Publishing, 2004); David S. Platt, *Why Software Sucks ... and What You Can Do about It* (Boston: Pearson Education, 2007).

6. Donald E. Knuth, "The Errors of T$_E$X," *Software—Practice & Experience* 19, no. 7 (July 1989): 610.

7. Andreas Zeller, *Why Programs Fail: A Guide to Systematic Debugging* (San Francisco: Morgan Kaufmann, 2006), 3–4. Zeller uses *infection* instead of the word *fault*, but I feel that the term *infection* is too reminiscent of illness in people, which makes bugs sound more subject to luck than they are.

8. "Storing the last two digits" generally referred to storing it on disk or tape, since most programs back then would be reading stored data, processing it, and writing the output back to storage; the programs would only store a two-character text string, such as "00" or "99"—the ASCII encoding (or whatever encoding was used) of the first digit followed by the encoding of the second digit, occupying two bytes total. These would then be converted, inside the program, into the corresponding integer before any mathematical operations were done.

9. Wikipedia, "Year 2000 Problem," accessed January 2, 2018, https://en.wikipedia .org/wiki/Year_2000_problem.

10. There are also a few other situations, such as dividing by 0, which will cause a program to crash, because there is no recoverable way to handle them; there is no correct value that can be stored as the result of a division by 0.

11. This code was widely reported; I copied it from Zune Boards, "Cause of Zune 30 Leapyear Problem ISOLATED!," December 31, 2008, accessed January 4, 2018, http:// www.zuneboards.com/forums/showthread.php?t=38143.

12. IBM, *DOS 1.00* (Boca Raton, FL: IBM, 1981), A-7.

13. Programs could instruct DOS that when it detected the problem, rather than call the code that presented the "Abort, Retry, Ignore" prompt, it should call code provided by the program.

14. Volvo, "Vision 2020," accessed January 4, 2018, http://www.volvocars.com/ intl/about/vision-2020.

15. Ole-Johan Dahl, Edsger W. Dijkstra, and C. A. R. Hoare, *Structured Programming* (London: Academic Press, 1972); Richard C. Linger, Harlan D. Mills, and Bernard I. Witt, *Structured Programming: Theory and Practice* (Reading, MA: Addison-Wesley, 1979).

16. Naur, Randell, and Buxton, *Software Engineering Concepts and Techniques*, 132–134, 292–295.

17. Donald E. Knuth, "Structured Programming with Go To Statements," *Computing Surveys* 6, no. 4 (December 1974): 261. The article gives a history of the battle over GOTO, and summarizes situations in which it is not needed, but also situations in which Knuth feels it makes the code cleaner.

18. Dahl, Dijkstra, and Hoare, *Structured Programming*, 1–2.

19. Linger, Mills, and Witt, *Structured Programming*, 1.

20. Ibid., 2.

21. Brian W. Kernighan and P. J. Plauger, *The Elements of Programming Style*, 2nd ed. (New York: McGraw-Hill, 1978), 2, 9.

22. Ibid., ix, xi.

23. Brian W. Kernighan and Dennis M. Ritchie, *The C Programming Language* (Englewood Cliffs, NJ: Prentice-Hall, 1978), 63.

24. Brian W. Kernighan and P. J. Plauger, *Software Tools* (Reading, MA: Addison-Wesley, 1976), 285. The book is about small tools that help programmers with their work, including a tool that lets them count how many times they different language keywords, so their statement about having no GOTOs is based on actually analyzing the code for all the tools they describe.

25. Gerald M. Weinberg, *The Psychology of Computer Programming*, silver anniversary ed. (New York: Dorset House, 1998), 45–46.

26. "The Calling of an Engineer," accessed January 4, 2018, https://www.camp1.ca/wordpress/?page_id=2.

27. Michael A. Cusumano and Richard W. Selby, *Microsoft Secrets: How the World's Most Powerful Software Company Creates Technology, Shapes Markets, and Manages People* (New York: Touchstone, 1998), 37, 43.

28. Dijkstra, quoted in Naur, Randell, and Buxton, *Software Engineering Concepts and Techniques*, 73; Edsger W. Dijkstra, "Structured Programming," in *Software Engineering Concepts and Techniques: Proceedings of the NATO Conferences*, ed. Peter Naur, Brian Randell, and J. N. Buxton (New York: Petrocelli/Charter, 1976), 223; Dahl, Dijkstra, and Hoare, *Structured Programming*, 6.

29. Linger, Mills, and Wiatt, *Structured Programming*, 13.

30. Cem Kaner, *Testing Computer Software* (Blue Ridge Summit, PA: Tab Books, 1988), 17, 21.

31. Glenford J. Myers, *The Art of Software Testing* (New York: John Wiley, 1979), 5–7.

32. Kaner, *Testing Computer Software*, 17.

33. Ibid., 24.

34. Myers, *Art of Software Testing*, 6.

35. Ibid., 12–13.

36. Cusumano and Selby, *Microsoft Secrets*, 40. I certainly experienced this too!

37. David Maynor, *Metasploit Toolkit* (Burlington, MA: Syngress Publishing, 2007), 218.

38. Ellen Ullman, *The Bug* (New York: Doubleday, 2003), 54–55.

39. Harlan D. Mills, "Software Development," in *Software Productivity* (New York: Dorset House, 1988), 243.

Chapter 6

1. Niklaus Wirth, *Algorithms + Data Structures = Programs* (Englewood Cliffs, NJ: Prentice-Hall, 1976).

2. David H. Ahl, ed., *Basic Computer Games: Microcomputer Edition* (New York: Workman Publishing, 1978), 89–90.

3. Wirth, *Algorithms + Data Structures = Programs*, 56, xiii.

4. Or store the structure in a global variable, which is usually a bad idea and makes the code harder to understand.

5. R. J. Pooley, *An Introduction to Programming in SIMULA* (Oxford: Blackwell Scientific, 1987), 167.

6. Ibid., 287.

7. Graham M. Birtwistle, Ole-Johan Dahl, Bjørn Myhrhaug, and Kristen Nygaard, *SIMULA Begin* (London: Input Two-Nine, 1973), 34.

8. Pooley, *Introduction to Programming in SIMULA*, 139.

9. Bjarne Stroustrup, *The Design and Evolution of C++* (Reading, MA: Addison-Wesley, 1994), 64.

10. Ibid., 44, 30, 94, 49.

11. Stein Krogdahl, "Concepts and Terminology in the Simula Programming Language," April 2010, 5, accessed January 25, 2018, http://folk.uio.no/simula67/Archive/concepts.pdf.

12. David Parnas, "The Secret History of Information Hiding," in *Software Pioneers: Contributions to Software Engineering*, ed. Manfred Broy and Ernst Denert (Berlin: Springer, 2002), 401.

13. David Parnas, "Information Distribution Aspects of Design Methodology," in *Information Processing 71, Proceedings of IFIP Congress 71, Volume 1—Foundations and*

Systems, ed. C. V. Freiman, John E. Griffith, and J. L. Rosenfeld (Amsterdam: North-Holland, 1972), 339, 340–341.

14. David Parnas, "On the Criteria to Be Used in Decomposing Systems into Modules," *Communications of the ACM* 15, no. 12 (December 1972): 1056.

15. Bjarne Stroustrup, *The C++ Programming Language* (Reading, MA: Addison-Wesley, 1986), iii.

16. Adele Goldberg and David Robson, *Smalltalk-80: The Language* (Reading, MA: Addison-Wesley, 1989), x.

17. Stroustrup, *Design and Evolution of C++*, 44.

18. Adapted from an example in Goldberg and Robson, *Smalltalk-80*, 34.

19. Ibid., 3, 7.

20. Stroustrup, *Design and Evolution of C++*, 4.

21. Brad J. Cox, *Object-Oriented Programming: An Evolutionary Approach* (Reading, MA: Addison-Wesley, 1986), iv, 51.

22. Bertrand Meyer, "Genericity vs. Inheritance" (paper presented at OOPSLA 1986, Portland, Oregon, September 29–October 2, 1986). In what is quite an interesting paper, Meyer explains that genericity is the notion of allowing, say, an array sort routine to be written in a generic way so that it can sort any type of data but still use the same sorting code. As Meyer points out, this idea, which was popularized in the language Ada, had not been discussed much among object-oriented theorists, although it turns out to be quite useful in certain situations and now exists in many object-oriented languages.

23. Bertrand Meyer, *Object-Oriented Software Construction* (New York: Prentice Hall, 1988), xiii.

24. Ibid.

25. Ibid., 41.

26. Birtwistle et al., *SIMULA Begin*, 209; Pooley, *Introduction to Programming in SIMULA*, 156.

27. Meyer, *Object-Oriented Software Construction*, 50.

28. Ibid., 376–379.

29. Linden J. Ball, Balder Onarheim, and Bo T. Christensen, "Design Requirements, Epistemic Uncertainty, and Solution Development Strategies in Software Design," in *Software Designers in Action: A Human-Centric Look at Design Work*, ed. André van der Hoek and Marian Petre (Boca Raton, FL: CRC Press, 2014), 232, 245. The study they referenced was Linden J. Ball, Jonathan St. B. T. Evans, Ian Dennis, and Thomas C.

Ormerod, "Problem-Solving Strategies and Expertise in Engineering Design," *Thinking and Reasoning* 3, no. 4 (1997): 247–270.

30. Joshua Bloch, "How to Design a Good API and Why It Matters" (paper presented at OOPSLA 2006, Portland, Oregon, October 22–26, 2006).

31. Henry Baird, interview with the author, August 7, 2017.

32. Stroustrup, *C++ Programming Language*, 8.

33. Meyer, *Object-Oriented Software Construction*, 49.

34. Gordon E. Moore, "Cramming More Components onto Integrated Circuits," *Electronics* 38, no. 8 (April 19, 1965): 114–117.

35. Dennis de Champeaux, Al Anderson, and Ed Feldhousen, "Case Study of Object-Oriented Software Development" (paper presented at OOPSLA 1992, Vancouver, British Columbia, October 18–22, 1992).

36. Frederick P. Brooks Jr., "No Silver Bullet—Essence and Accident in Software Engineering," in *The Mythical Man-Month: Essays on Software Engineering*, anniversary ed. (Boston: Addison-Wesley, 1995), 181–182.

37. Robert L. Grady, *Successful Software Process Improvement* (Upper Saddle River, NJ: Prentice Hall, 1997), 185–186.

38. Karl J. Lieberherr, Ian Holland, and Arthur J. Riel, "Object-Oriented Programming: An Objective Sense of Style" (paper presented at OOPSLA 1988, San Diego, California, September 25–30, 1988).

39. Ibid.

40. Mary Beth Rosson and Eric Gold, "Problem-Solution Mapping in Object-Oriented Design" (paper presented at OOPSLA 1989, New Orleans, Louisiana, October 2–6, 1989).

41. John A. Lewis, Sallie M. Henry, Dennis G. Kufara, and Robert S. Schulman, "An Empirical Study of the Object-Oriented Paradigm and Software Reuse" (paper presented at OOPSLA 1991, Phoenix, Arizona, October 6–11, 1991) (emphasis added).

42. The group at Virginia Tech did a separate study on maintainability in which they compared Objective-C and C. Sallie Henry, Matthew Humphrey, and John Lewis, "Evaluation of the Maintainability of Object-Oriented Software" (paper presented at the IEEE Region 10 Conference on Computer and Communication Systems, Hong Kong, September 1990).

43. Marvin V. Zelkowitz, "Education of Software Engineers," in *Perspectives on the Future of Software Engineering: Essays in Honor of Dieter Rombach*, ed. Jürgen Münch and Klaus Schmid (Berlin: Springer, 2013), 356.

44. Rick DeNatale, Charles Irby, John LaLonde, Burton Leathers, and Reed Phillips, "OOP in the Real World" (panel at OOPSLA/ECOOP 1990, Ottawa, Canada, October 21–25, 1990).

45. Cox, *Object-Oriented Programming*, 2.

46. In fact, there was a long-running argument inside Microsoft about whether object-oriented public API interfaces caused independent layers to be bound together too tightly and therefore were inferior to procedural ones.

47. Lucy Berlin, "When Objects Collide: Experiences with Reusing Multiple Class Hierarchies" (paper presented at OOPSLA/ECOOP 1990, Ottawa, Canada, October 21–25, 1990).

48. Neal Stephenson, *In the Beginning ... Was the Command Line* (New York: William Morrow, 1999).

49. Brian W. Kernighan and Rob Pike, *The UNIX Programming Environment* (Englewood Cliffs, NJ: Prentice-Hall, 1984), viii.

50. Cox, *Object Oriented Programming*, 18.

51. Meyer, *Object-Oriented Software Construction*, 14.

Chapter 7

1. In some languages, through *private inheritance*, the `Emailer` class could avoid exposing the `AESEncrypter` class to callers, but the code inside the `Emailer` class could still access members of the `AESEncrypter` class and in general be more tightly coupled to it (and therefore more likely to break due to internal changes in `AESEncrypter`) than it needed to be, when all it needed was access to the encryption functionality.

2. Not too unexpected in the world of cryptography, where encryption algorithms once considered secure often turn out to be breakable once computers get more powerful.

3. Bertrand Meyer, *Object-Oriented Software Construction* (New York: Prentice Hall, 1988), 242. He does give a few examples where he feels concrete multiple inheritance is a good idea, but none are particularly convincing.

4. Erich Gamma, Richard Helm, Ralph Johnson, and John Vlissides, *Design Patterns: Elements of Reusable Object-Oriented Software* (Boston: Addison-Wesley, 1995).

5. Christopher Alexander, Sara Ishikawa, and Murray Silverstein, with Max Jacobson, Ingrid Fiksdahl-King, and Shlomo Angel, *A Pattern Language: Towns · Building · Construction* (New York: Oxford University Press, 1977).

6. Ibid., xix–xxxiv. That is a list of the patterns; the bulk of the book, over a thousand pages, is a detailed discussion of each one.

7. Walter Tichy, "The Evidence for Design Patterns," in *Making Software: What Really Works, and Why We Believe It*, ed. Andy Oram and Greg Wilson (Sebastopol, CA: O'Reilly, 2011), 393–414.

8. Gamma et al., *Design Patterns*, 20, 18.

9. Erich Gamma, "Design Patterns—Ten Years Later," in *Software Pioneers: Contributions to Software Engineering*, ed. Manfred Broy and Ernst Denert (Berlin: Springer, 2002), 692.

10. Kent Beck, "Simple Smalltalk Testing: With Patterns," accessed January 8, 2018, http://swing.fit.cvut.cz/projects/stx/doc/online/english/tools/misc/testfram.htm.

11. Michael C. Feathers, *Working Effectively with Legacy Code* (Upper Saddle River, NJ: Prentice Hall, 2004), 13.

12. Alan Shalloway and James R. Trott, *Design Patterns Explained: A New Perspective on Object-Oriented Design*, 2nd ed. (Boston: Addison-Wesley, 2005), 352–353, 400.

13. Bill Blunden, *Software Exorcism: A Handbook for Debugging and Optimizing Legacy Code* (Berkeley, CA: Apress, 2003).

14. Feathers, *Working Effectively with Legacy Code*, xvi.

15. Joel Spolsky, "Don't Let Architecture Astronauts Scare You," *Joel on Software* blog, April 21, 2001, accessed January 8, 2018, https://www.joelonsoftware.com/2001/04/21/dont-let-architecture-astronauts-scare-you/.

16. Ibid.

17. Antony Tang, Aldeida Aleti, Janet Burge, and Hans van Vliet, "What Makes Software Design Effective," in *Software Designers in Action: A Human-Centric Look at Design Work*, ed. André van der Hoek and Marian Petre (Boca Raton, FL: CRC Press, 2014), 134.

18. Robert Martin, ed., *Clean Code: A Handbook of Agile Software Craftmanship* (Upper Saddle River, NJ: Prentice Hall, 2009); Andrew Hunt and David Thomas, *The Pragmatic Programmer: From Journeyman to Master* (Boston: Addison-Wesley, 2000).

19. Atul Gawande, *The Checklist Manifesto: How to Get Things Right* (New York: Metropolitan Books, 1999).

20. Gerald M. Weinberg, *The Psychology of Computer Programming*, silver anniversary ed. (New York: Dorset House, 1998), 16.

21. Ibid., 21.

22. Ole-Johan Dahl, Edsger W. Dijkstra, and C. A. R. Hoare, *Structured Programming* (London: Academic Press, 1972), 6.

23. Jon Bentley, *Writing Efficient Programs* (Englewood Cliffs, NJ: Prentice-Hall, 1982), xii.

24. Donald E. Knuth, "Structured Programming with Go To Statements," *Computing Surveys* 6, no. 4 (December 1974): 268.

25. Wikipedia, "Year 2038 Problem," accessed January 9, 2018, https://en.wikipedia.org/wiki/Year_2038_problem.

26. William Porquet, "The Project 2038 Frequently Asked Questions (FAQ)," August 15, 2007, accessed January 9, 2018, http://maul.deepsky.com/~merovech/2038.html. This FAQ shows the effect of the second clock ticking over; it also lists operating systems that have proactively resolved the problem.

Chapter 8

1. Wikipedia, "Code Red (Computer Worm)," accessed January 9, 2018, https://en.wikipedia.org/wiki/Code_Red_(computer_worm).

2. Worm writers frequently look at what code is fixed in a patch as a way to locate exploitable code, since not everybody applies patches right away.

3. eWeek Editors, "Microsoft: XP Dramatically More Secure," October 22, 2001, accessed January 9, 2018, http://www.eweek.com/news/microsoft-xp-dramatically-more-secure.

4. Wikipedia, "Timeline of Computer Viruses and Worms," accessed January 9, 2018, https://en.wikipedia.org/wiki/Timeline_of_computer_viruses_and_worms.

5. The amount of data to be copied is usually specified by a variable, whose value won't be known until the code is running, but with clever analysis you can often figure out what range of values the variables might have. The primary tool is Structured Annotation Language, which is now shipping with Microsoft's compiler products. Excuse the advertisement for Microsoft, but it really is a worthwhile attempt to fix the problem, and for everybody's code, not just Microsoft's.

6. Eric Limer, "How Heartbleed Works: The Code behind the Internet's Security Nightmare," accessed January 9, 2018, http://gizmodo.com/how-heartbleed-works-the-code-behind-the-internets-se-1561341209.

7. Nadine Kano, *Developing International Software for Windows 95 and Windows NT* (Redmond, WA: Microsoft Press, 1995), 488.

8. It was also faster to display characters than graphics. Here's a fun fact: if you did have a graphics display, the actual glyphs for the upper 128 characters displayed by

MS-DOS were stored in memory and could be overwritten, so that any given character in the 128 to 255 range could show any arbitrary arrangement of dots that fit in an 8 by 8 grid. The game *Microsoft Decathlon*, which came out in 1982, used this trick to display "graphics" of athletes using characters instead of full graphics; it stored every possible body part needed in one of the upper 128 characters.

9. Kano, *Developing International Software*, 496.

10. This chart and the next one were produced using a BASIC program I wrote to print them out, running on the amazing PC-BASIC interpreter written by Rob Hagemans. "PC-BASIC," accessed January 9, 2018, http://www.pc-basic.org.

11. Kano, *Developing International Software*. In addition to going into detail on how this all works, the book includes many pages of sumptuous character glyphs for you to gaze at.

12. Ibid., 464. This is the Windows equivalent of the MS-DOS Latin US code page, which was called Latin 1 or ANSI. It removes the line-drawing characters and Greek letters, adds more symbols (such as opening and closing single and double quotes, which were absent in MS-DOS, so programs could only use the generic single and double quotes that exist in standard ASCII), and has a more complete collection of accented letters. Not the Turkish *I*, though; that still required a Turkish code page.

13. Unicode has different encodings, and although early versions of Windows NT really did only support exactly 2 bytes per character in all cases (an encoding known as UCS-2), Unicode eventually grew to encompass more than the 65,536 characters that could fit in even 16 bits, so Windows switched to an encoding called UTF-16, in which *most* characters were encoded in 2 bytes, but certain characters were encoded in 4 bytes. I will ignore it and treat Unicode as if every character is 2 bytes, which is what the code in question does. In particular, in UTF-16 the characters encoded in 4 bytes were set up so that neither of the 2-byte values (known as the high and low surrogate) was a valid 2-byte character—so since the code I am discussing here was mostly copying strings around, or might occasionally be searching for a common character like a period or backslash that existed in the lower half of the original ASCII table (whose encoding in UTF-16 was equal to the 16-bit version of that ASCII value and therefore always stored in 2 bytes, not 4), it would work fine, completely oblivious to the UCS-2 versus UTF-16 difference (as `wcslen()` also was).

14. Unicode, Inc., "Full Emoji List v5.0," accessed January 10, 2018, https://unicode. org/emoji/charts/full-emoji-list.html. Since you asked, the "pile of poo" emoji is assigned the value 1F4A9 (that is, a base 16, aka hexadecimal, number, equal to 128,169 in decimal). C# is also Unicode only (it uses UTF-16 encoding), and hence the "Turkish *I*" problem discussed in chapter 3 can manifest itself. The Turkish *I* problem with upper- and lowercasing of strings is not related to being able to fit the characters in the Unicode character set; it's just about the algorithm used to convert uppercase to lowercase. In the days of ASCII, the mapping was set up for the

26 English letters, so the lowercase version of any letter was always 32 above the uppercase letter—for example, *A* was 65 and *a* was 97—and conversion between cases was a simple mathematical operation. In the world of code pages *or* Unicode, it has to be done by having a table with explicit mapping for every character, and which table you use depends on the locale the computer is set up to use; for upper-casing, the English mapping will map *i* to *I*, and the Turkish mapping will map *i* to *İ*.

15. Some people feel the way it was integrated into C/C++ on Windows was poorly done, though, because while it doesn't matter for the low-level code discussed here, for user interface (UI) code the fact that UTF-16 can sometimes encode a character in 4 byes instead of 2 can be important. See "UTF-8 Everywhere," accessed January 10, 2018, http://utf8everywhere.org/ for details.

16. Especially given that in the encoding system known as *multibyte characters*, another system used for encoding ideographic alphabets in which characters may be encoded in 1, 2, or 3 bytes, the corresponding API `mbslen()` had already been defined to return a character count, not a byte count. Multibyte characters, in a simplified subset known as double-byte characters, were the source of the notorious Microsoft interview-coding question "Kanji backspace." Having the solution published in a book didn't dissuade Microsoft interviewers from continuing to ask it. Kano, *Developing International Software*, 70.

17. Specifically, it is not worth explaining because I'm not actually sure why it is there, although it is related to a fashion in vogue at the time of appending `_t` to type-related constructs. In writing this footnote, I am subscribing to the Internet theory, "If you want somebody to explain something to you, then give a wrong explanation in public." Also, on some systems, `wchar_t` is 32 bits, not 16 bits—a fact that is not relevant to the discussion here.

18. You could also write `strlen(other_sb_buffer) < sizeof(sb_buffer)`, which is what people likely wrote, since avoiding the subtraction of 1 makes the code slightly faster, but I find that conceptually a bit harder to read. These days, the compiler likely notices the combination of `<=` and `- 1`, and converts it to the faster code anyway.

19. Technically it is cleaner to divide by `sizeof(wc_buffer[0])` rather than by `sizeof(wchar_t)`, in case the type changes, but in this instance that is unlikely to matter; it is more important in code that is meant to compile, through a few tricks, for both single- and double-byte characters, but Windows internally was entirely double byte. Microsoft's C compiler also has a pseudo-API called `_countof()`, which will count array elements properly, but that is not standard in other C compilers.

20. Microsoft did define such a type, known as `BSTR`, but it *was* more complicated to use, and since it had to interoperate with various APIs that expected plain 0-terminated strings, it could not be required everywhere. Microsoft, "BSTR," accessed

January 10, 2018, https://msdn.microsoft.com/en-us/library/windows/desktop/ms
221069(v=vs.85).aspx.

21. This example required overloading the = operator as well. Operator overloading
originally appeared in the language Algol 68, which allowed operators to be named
using any combination of letters and symbols. A few of the members of the committee that designed Algol 68, including Dijkstra and Hoare, filed a dissenting opinion
stating that the language was too complicated, although I don't know if operator
overloading was the straw that broke the camel's back. Edsger W. Dijkstra, Fraser
Duncan, Jan Garwick, C. A. R. Hoare, Brian Randell, Gerhard Seegmueller, Wlad
Turski, and Michael Woodger, "Minority Report," accessed January 5, 2018, http://
archive.computerhistory.org/resources/text/algol/algol_bulletin/A31/P111.HTM.

22. They specify, respectively, the access mode, sharing mode, security attributes,
creation disposition (what to do if the file doesn't exist), flags and attributes (various
options such as whether the file should be encrypted or read only), and template file
(another file whose attributes should be reused for this file). Microsoft, "CreateFile
Function," accessed January 10, 2018, https://msdn.microsoft.com/en-us/library/
windows/desktop/aa363858(v=vs.85).aspx.

23. A few points for those who are learning Windows programming from my examples. For one, the reason that the operation is split into three—open, write, and close
versus having WriteFile() take the filename as a parameter and do the entire
operation inside one API call—is mostly for performance, in that it is faster to open
the file once and then do multiple writes to the open handle, just as if you are going
to insert multiple pages into a file folder, it is easier to find it in the filing cabinet
once and then leave it on your desk while you insert each page, as opposed to finding it in the filing cabinet (the CreateFile()) and then returning it to the filing
cabinet (the CloseHandle()) around each page you insert. Second, the Close-
Handle() is not strictly needed right here, in that Windows will close open handles
when the program exits, but it makes the data available to other programs on the
computer at that point, which might be useful—just as you could keep any physical
files you had used that day on your desk until you were about to leave for the day
and then put them all back, but then if somebody else went to the filing cabinet
looking for one of those files, they wouldn't find it.

24. Microsoft, "CreateFile Function."

25. In most modern languages, the indenting doesn't affect how the compiler inter-
prets the code, although there are exceptions, most notably the language Python.

26. ERROR_ARENA_TRASHED is not explained in my old DOS manuals. If you search
the Internet you will see it explained with the opaque message, "The storage control
blocks were destroyed." Private Facebook discussion with a few Microsoft old-timers,
while unsuccessful in producing an admission that the name was chosen because it
sounded amusing, revealed that DOS would return it when it thought that a pro-

gram had written to memory that wasn't assigned to it. If you have installed Visual Studio Community per my footnoted instructions in chapter 1, you can observe the error "in the wild" lurking in the file `Error.h` (which states that it contains DOS error codes) in a section labeled "These are the 2.0 error codes," between `ERROR_INVALID_HANDLE` and `ERROR_NOT_ENOUGH_MEMORY`. Microsoft Corporation, "Error.h," installed at `C:\Program Files (x86)\Microsoft SDKs\Windows\v7.1A\Include\Error.h`.

27. Yes, I am skipping some details about when the `FileStream` object gets finalized.

28. Microsoft, "Unchecked (C# Reference)," accessed January 10, 2018, https://msdn.microsoft.com/en-us/library/a569z7k8.aspx.

29. Bjarne Stroustrup, *Design and Evolution of C++* (Reading, MA: Addison-Wesley, 1994), 383, 126, 128.

30. Bjarne Stroustrup, "A History of C++: 1979–1991," *ACM SIGPLAN Notices* 28, no. 3 (March 1993): 271–297.

31. "The danger in trying to force object-oriented concepts onto a C base is to get an inconsistent construction, impairing the software development process and the quality of the resultant products. A hybrid approach yields hybrid quality. This is why serious reservations may be voiced about the object-oriented extensions of C [which include C++]." Bertrand Meyer, *Object-Oriented Software Construction* (New York: Prentice Hall, 1988), 382.

32. IBM, "PL/I Condition Handling Semantics," accessed January 11, 2018, https://www.ibm.com/support/knowledgecenter/SSLTBW_2.1.0/com.ibm.zos.v2r1.ceea200/pliax.htm; John Goodenough, "Structured Exception Handling" (paper presented at Principles of Programming Languages 1975, Palo Alto, California, January 20–22, 1975).

33. Stroustrup, *Design and Evolution of C++*, 270.

34. "Bjarne Stroustrup's FAQ," accessed January 11, 2018, http://www.stroustrup.com/bs_faq.html. He confirms this quote in the "Did You Really Say That?" section of his FAQ; he also says he doubts he came up with the quote originally.

35. Peter Seibel, *Coders at Work: Reflections on the Craft of Programming* (New York: Apress, 2009), 163, 170, 193, 224.

36. Ibid., 10, 63, 475.

37. Robert Merkel, "How the Heartbleed Bug Reveals a Flaw in Online Security," *The Conversation*, April 11, 2014, accessed January 28, 2018, https://theconversation.com/how-the-heartbleed-bug-reveals-a-flaw-in-online-security-25536.

38. I looked at the top-ten departments according to an Internet ranking: Carnegie Mellon University, http://coursecatalog.web.cmu.edu/schoolofcomputerscience/; University of California at Berkeley, http://guide.berkeley.edu/departments/ electrical-engineering-computer-sciences/#computer-science; Princeton University, https://www.cs.princeton.edu/courses/catalog; Stanford University, https://cs .stanford.edu/academics/courses; Massachusetts Institute of Technology, http:// catalog.mit.edu/degree-charts/computer-science-engineering-course-6-3/; University of Illinois at Urbana-Champaign, http://cs.illinois.edu/academics/courses; Cornell University, https://www.cs.cornell.edu/courseinfo/ListofCSCourses/; University of Washington, http://www.cs.washington.edu/education/courses/; Georgia Tech, http://www.cc.gatech.edu/three-year-course-outline; University of Texas at Austin, https://login.cs.utexas.edu/undergraduate-program/academics/curriculum/courses. Berkeley has specific courses in C, Scheme, C++, Java, and Python. Stanford teaches Python and JavaScript. MIT uses Python for its introductory class. Cornell teaches Python and C++. And Texas teaches C++.

39. The Joint Task Force on Computing Curricula, "Computer Science Curricula 2013," accessed January 11, 2018, http://cs2013.org.

40. Robert Harper, *Practical Foundations for Programming Languages*, 2nd ed. (New York: Cambridge University Press, 2016).

41. Joint Task Force on Computing Curricula, "Computer Science Curricula 2013," 380.

42. Robert Harper, interview with the author, May 16, 2017.

43. Rob Pike, "The Good, the Bad, and the Ugly: The Unix™ Legacy" (presentation at uptime(1) conference, Copenhagen, September 8–9, 2001), accessed January 14, 2018, http://www.herpolhode.com/rob/ugly.pdf.

44. Cristina Videira Lopes, *Exercises in Programming Style* (Boca Raton, FL: CRC Press, 2014), xii.

45. Raymond Queneau, *Exercises in Style*, trans. Barbara Wright (New York: New Directions, 1981).

46. Henry Baird, interview with the author, August 7, 2017.

47. Joint Task Force on Computing Curricula, "Computer Science Curricula 2013," 381.

48. Doug Sheppard, "Beginner's Introduction to Perl," October 16, 2000, accessed January 11, 2018, https://www.perl.com/pub/2000/10/begperl1.html.

49. Niklaus Wirth, "A Brief History of Software Engineering," *IEEE Annals of the History of Computing* 30, no. 3 (July–September 2008): 38–39.

Chapter 9

1. Mark R. Milligan, "How Was Utah's Topography Formed?" accessed January 11, 2018, http://geology.utah.gov/surveynotes/gladasked/gladtopoform.htm.

2. Mitch Lacey, "The History of the Agile Manifesto" (undated blog entry), accessed January 11, 2018, http://www.mitchlacey.com/blog/the-history-of-the-agile-mani festo.

3. G. Pascal Zachary, *Showstopper: The Breakneck Race to Create Windows NT and the Next Generation at Microsoft* (New York: Free Press, 1994), 65. The first official planned completion date was March 1991; at the beginning of 1992, it was hoped that it would ship by June 1992 (ibid., 177). It eventually shipped in summer 1993. At the beginning of the project, in late 1988, the first estimate was that the entire thing would take eighteen months.

4. "Manifesto for Agile Software Development," accessed January 12, 2018, http://agilemanifesto.org/.

5. Jeff Sutherland, "Business Object Design and Implementation Workshop" (workshop at OOPSLA 1995, Austin, Texas, October 15–19, 1995). Sutherland was the chair of the workshop; although the original paper lists only Schwaber as the author (Ken Schwaber, "Scrum Development Process," accessed January 12, 2018, http://www.jeffsutherland.org/oopsla/schwaber.html), the "Scrum Guide," written later by Schwaber and Sutherland, states that they copresented it (Jeff Sutherland and Ken Schwaber, "The Scrum Guide," accessed January 12, 2018, http://www.scrumguides .org/docs/scrumguide/v2016/2016-Scrum-Guide-US.pdf).

6. Sutherland and Schwaber, "Scrum Guide."

7. Martin Fowler, foreword to *Agile Software Development with Scrum*, by Ken Schwaber and Mike Beedle (Upper Saddle River, NJ: Prentice Hall, 2002), vi.

8. Harlan D. Mills, "In Retrospect," in *Software Productivity* (New York: Dorset House, 1988), 2.

9. Frederick P. Brooks Jr., "The Mythical Man-Month," in *The Mythical Man-Month: Essays on Software Engineering*, anniversary ed. (Boston: Addison-Wesley, 1995), 20.

10. Frederick P. Brooks Jr., *"The Mythical Man-Month* after 20 Years," in *The Mythical Man-Month: Essays on Software Engineering*, anniversary ed. (Boston: Addison-Wesley, 1995), 264.

11. Frederick P. Brooks Jr., "Plan to Throw One Away," in *The Mythical Man-Month: Essays on Software Engineering*, anniversary ed. (Boston: Addison-Wesley, 1995), 116.

12. Brooks, *"Mythical Man-Month* after 20 Years," 265.

13. Ibid., 266.

14. Ibid., 266.

15. Ibid., 268.

16. Sutherland, "Business Object Design and Implementation Workshop," 170, 174.

17. Ibid., 174. The term was capitalized at this stage.

18. Cusumano and Selby, *Microsoft Secrets*, 263, 276, 294.

19. Brooks, "*Mythical Man-Month* after 20 Years," 270.

20. "The Rules of Extreme Programming," accessed January 12, 2018, http://www
.extremeprogramming.org/rules.html.

21. Schwaber, "Scrum Development Process."

22. Kent Beck, *Extreme Programming Explained: Embrace Change* (Boston: Addison-Wesley, 2000), 157, 158.

23. Ken Schwaber and Mike Beedle, *Agile Software Development with Scrum* (Upper Saddle River, NJ: Prentice Hall, 2002), 57–58.

24. Schwaber, "Scrum Development Process."

25. Harlan D. Mills, "Software Engineering Education," in *Software Productivity* (New York: Dorset House, 1988), 253.

26. Schwaber and Beedle, *Agile Software Development with Scrum*, 2, 109.

27. Beck, *Extreme Programming Explained*, 3, 4.

28. Plato, "Apology," in *Plato: Complete Works*, ed. John M. Cooper, trans. G. M. A. Grube (Indianapolis: Hackett Publishing, 1997), 21, 22.

29. Watts S. Humphrey, *A Discipline for Software Engineering* (Boston: Addison-Wesley, 1995), x.

30. Humphrey, *Discipline for Software Engineering*, 8, 471–486.

31. Michael Fagan, "Design and Code Inspections to Reduce Errors in Program Development," *IBM Systems Journal* 15, no. 3 (1976): 182–211.

32. Tom Gilb and Dorothy Graham, *Software Inspection* (Harlow, UK: Addison-Wesley, 1993), 33–36.

33. Trevor Reese, "Implementing Document Inspection on an Unusually Wide Basis at an Electronics Manufacturer," in Tom Gilb and Dorothy Graham, *Software Inspection* (Harlow, UK: Addison-Wesley, 1993), 307.

34. Humphrey, *Discipline for Software Engineering*, 365–368.

35. Watts S. Humphrey, *Introduction to the Team Software Process*[SM] (Reading, MA: Addison-Wesley, 2000).

36. Humphrey, *Discipline for Software Engineering*, 262.

37. Ibid., 263–265.

38. Steve McConnell, *Software Estimation: Demystifying the Black Art* (Redmond, WA: Microsoft Press, 2006), 35–41.

39. Ibid., 149–155.

40. Tracy Kidder, *The Soul of a New Machine* (New York: Modern Library, 1981), 82–83.

41. Ofer Sharone, "Engineering Overwork: Bell-Curve Management at a High-Tech Firm," in *Fighting For Time: Shifting Boundaries of Work and Social Life*, ed. Cynthia Fuchs Epstein and Arne L. Kalleberg (New York: Russell Sage Foundation, 2004), 192.

42. Kidder, *Soul of a New Machine*, 152–153.

43. Schwaber and Beedle, *Agile Software Development with Scrum*, 35.

44. Beck, *Extreme Programming Explained*, 60.

45. Mary Shaw, "Prospects for an Engineering Discipline," *IEEE Software* 7, no. 6 (November 1990): 21.

46. Richard C. Linger, Harlan D. Mills, and Bernard I. Witt, *Structured Programming: Theory and Practice* (Reading, MA: Addison-Wesley, 1979), 147.

47. Harlan D. Mills, "Reading Code as a Management Activity," in *Software Productivity* (New York: Dorset House, 1988), 179–184; Harlan D. Mills, "The Case against GO TO Statements in PL/I," in *Software Productivity* (New York: Dorset House, 1988), 27

48. Gerald M. Weinberg, *The Psychology of Computer Programming*, silver anniversary ed. (New York: Dorset House, 1998), 5–14.

49. Adam Barr, *Find the Bug: A Book of Incorrect Programs* (Boston: Addison-Wesley, 2005), 3–32.

50. Michael A. Cusumano and Richard W. Selby, *Microsoft Secrets: How the World's Most Powerful Software Company Creates Technology, Shapes Markets, and Manages People* (New York: Touchstone, 1998), 288.

51. Beck, *Extreme Programming Explained*, 110.

52. Schwaber and Beedle, *Agile Software Development with Scrum*, viii.

53. Scott Rosenberg, *Dreaming in Code: Two Dozen Programmers, Three Years, 4732 Bugs, and One Quest for Transcendent Software* (New York: Crown Publishers, 2007), 239, 264.

54. Ivar Jacobson, Pan-Wei Ng, Paul E. McMahon, Ian Spence, and Svante Lidman, *The Essence of Software Engineering: Applying the SEMAT Kernel* (Upper Saddle River, NJ: Addison-Wesley, 2013), xxix, xxviii.

55. Greg Wilson, "Two Solitudes" (keynote talk at the SPLASH 2013 conference, Indianapolis, October 26–31, 2013).

56. Geary A. Rummler, *Serious Performance Consulting according to Rummler* (San Francisco: Pfeiffer, 2007), xiii.

57. Schwaber and Beedle, *Agile Software Development with Scrum*, 5.

58. "It is tempting to imagine the Big Bang to be like the beginning of a concert. You're seated for a while fiddling with your program, and then suddenly at $t = 0$ the music starts. But the analogy is mistaken. ... [T]he singularity at the beginning of the universe is not an event in time. Rather, it is a temporal boundary or edge. There are no moments of time 'before' $t = 0$. ... As Grünbaum is fond of saying, even though the universe is finite in age, it has always existed, if by 'always' you mean at all instants of time." Jim Holt, *Why Does the World Exist?* (London: W. W. Norton, 2012), 75.

Chapter 10

1. Randall Rustin, ed., *Debugging Techniques in Large Systems* (Englewood Cliffs, NJ: Prentice-Hall, 1971).

2. IEEE Computer Society, "Harlan D. Mills Award," accessed January 13, 2018, https://www.computer.org/web/awards/mills.

3. Harold Sackman, *Man-Computer Problem Solving* (Princeton, NJ: Auerbach Publishers, 1970).

4. Maurice H. Halstead, *Elements of Software Science* (New York: North Holland, 1977).

5. Victor R. Basili and Harlan D. Mills, "Understanding and Documenting Programs," *IEEE Transactions on Software Engineering* 8, no. 3 (May 1982): 270–283.

6. Ralph T. Putnam, D. Sleeman, Juliet A. Baxter, and Laiani K. Kuspa, "A Summary of Misconceptions of High-School BASIC Programmers," in *Studying the Novice Programmer*, ed. Elliott Soloway and James C. Spohrer (Hillsdale, NJ: Lawrence Erlbaum Associates, 1989).

7. Ben Schneiderman, *Software Psychology: Human Factors in Computer and Information Systems* (Boston: Little, Brown, 1980), 66–70.

8. Larry Weissman, "Psychological Complexity of Computer Programs: An Experimental Methodology," *ACM SIGPLAN Notices* 9, no. 6 (June 1974): 25–36.

9. Tom Love, "An Experimental Investigation of the Effect of Program Structure on Program Understanding," *ACM SIGSOFT Software Engineering Notes* 2, no. 2 (March 1977): 105–113; Schneiderman, *Software Psychology*, 72–74.

10. Max E. Sime, Thomas R. G. Green, and D. John Guest, "Psychological Evaluation of Two Conditional Constructions Used in Computer Languages," *International Journal of Man-Machine Studies* 5, no. 1 (1973): 105–113; Henry C. Lucas Jr. and Robert B. Kaplan, "A Structured Programming Experiment," *Computer Journal* 19, no. 2 (1976): 136–138.

11. Marilyn Bohl, *Flowcharting Techniques* (Chicago: Science Research Associates, 1971).

12. "donkey.bas," accessed January 13, 2018, https://github.com/coding-horror/donkey.bas/blob/master/donkey.bas.

13. Richard E. Mayer, "Different Problem-Solving Competencies Established in Learning Computer Programming with and without Meaningful Models," *Journal of Education Psychology* 67, no. 6 (1975): 725–734; Schneiderman, *Software Psychology*, 81–85.

14. Frederick P. Brooks Jr., "The Flow-Chart Curse," in *The Mythical Man-Month: Essays on Software Engineering*, anniversary ed. (Boston: Addison-Wesley, 1995), 167–168.

15. As my father put it, "In my career with over a hundred papers, there is something truly new (having no obvious antecedents) in only one of them. And I conjecture that is one more than average."

16. Gerald M. Weinberg, *The Psychology of Computer Programming*, silver anniversary ed. (New York: Dorset House, 1998), 6.

17. John Buxton and Brian Randell, introduction to "Part II: Report on a Conference Sponsored by the NATO Science Committee, Rome Italy, October 27–31, 1969," in *Software Engineering Concepts and Techniques*, ed. Peter Naur, Brian Randell, and J. N. Buxton (New York: Petrocelli/Charter, 1976), 145.

18. Roger M. Needham and Joel D. Aron, "Software Engineering and Computer Science," in *Software Engineering Concepts and Techniques*, ed. Peter Naur, Brian Randell, and J. N. Buxton (New York: Petrocelli/Charter, 1976), 251.

19. Christopher Strachey, quoted in "Theory and Practice," in *Software Engineering Concepts and Techniques*, ed. Peter Naur, Brian Randell, and J. N. Buxton (New York: Petrocelli/Charter, 1976), 147.

20. Wikipedia, "Sturgeon's Law," accessed January 13, 2018, https://en.wikipedia.org/wiki/Sturgeon%27s_law.

21. Donald Knuth, interview with the author, February 10, 2017.

22. Victor Basili, interview with the author, December 7, 2016.

23. Weinberg, *Psychology of Computer Programming*, 202.

24. Ben Shneiderman, "No Members, No Officers, No Dues: A Ten Year History of the Software Psychology Society," *ACM SIGCHI Bulletin* 18, no. 2 (October 1986): 14–16.

25. Ben Shneiderman, interview with the author, December 2, 2016.

26. National Museum of American History, "Microsoft Windows NT OS2 Design Workbook," no. 2001.3014.01, accessed January 13, 2018, http://americanhistory.si.edu/collections/search/object/nmah_742559.

27. Greg Wilson, "Two Solitudes" (keynote talk at the SPLASH 2013 conference, Indianapolis, October 26–31, 2013).

28. Steve McConnell, *Code Complete: A Practical Handbook of Software Construction* (Redmond, WA: Microsoft Press, 1993), 93–94.

29. Ibid., 202–206.

30. Steve McConnell, *Code Complete: A Practical Handbook of Software Construction*, 2nd ed. (Redmond, WA: Microsoft Press, 2004), 279–281.

31. Pierre Bourque and Richard E. Fairly, eds., *SWEBOK V3.0: Guide to the Software Engineering Body of Knowledge* (Piscataway, NJ: IEEE Computer Society, 2014); John White and Barbara Simons, "ACM's Position on the Licensing of Software Engineers," *Communications of the ACM* 45, no. 11 (November 2002): 91.

32. Bourque and Fairly, *SWEBOK V3.0*, 3–8.

33. Ibid., xxxii.

34. Paul Clements, Felix Bachmann, Len Bass, David Garlan, James Ivers, Reed Little, Paulo Merson, Robert Nord, and Judith Stafford, *Documenting Software Architectures: Views and Beyond*, 2nd ed. (Upper Saddle River, NJ: Addison-Wesley, 2011).

35. Richard Hall Thayer and Merlin Dorfman, eds., *Software Engineering Essentials, Volume 1: The Development Process* (Carmichael, CA: Software Management Training Press, 2013), 140.

36. Richard Johnson, "Object-Oriented Systems Development: A Review of Empirical Research," *Communications of the Association for Information System* 8 (2002): 65–81.

37. Barry Boehm, Hans Dieter Rombach, and Marvin V. Zelkowitz, eds., *Foundations of Empirical Software Engineering: The Legacy of Victor R. Basili* (Berlin: Springer, 2005).

Chapter 11

1. The book *Showstopper* describes stress tests. G. Pascal Zachary, *Showstopper: The Breakneck Race to Create Windows NT and the Next Generation at Microsoft* (New York: Free Press, 1994), 154–157.

2. An Internet image search for Ship-It Award reveals many examples.

3. The e-mail announcing the program accidentally referred to it as the Shit-It Award.

4. Frederick P. Brooks Jr., "Calling the Shot," in *The Mythical Man-Month: Essays on Software Engineering*, anniversary ed. (Boston: Addison-Wesley, 1995), 87. The quote is sometimes reported as "experience keeps a dear school ..."

5. Atul Gawande, *The Checklist Manifesto: How to Get Things Right* (New York: Metropolitan Books, 1999), 58.

6. Ibid., 161.

7. Ibid., 8–11.

8. Ibid., 11.

9. Frederick P. Brooks Jr., "No Silver Bullet—Essence and Accident in Software Engineering," in *The Mythical Man-Month: Essays on Software Engineering*, anniversary ed. (Boston: Addison-Wesley, 1995), 181.

10. Frederick P. Brooks Jr., "'No Silver Bullet' Refired," in *The Mythical Man-Month: Essays on Software Engineering*, anniversary ed. (Boston: Addison-Wesley, 1995), 208.

11. Robert L. Glass, "Glass" (column), *System Development* (January 1988): 4–5.

12. David Parnas, e-mail to the author, May 22, 2017.

13. Harlan Mills, foreword to *Fortran with Style: Programming Proverbs*, by Henry F. Ledgard and Louis J. Chmura (Rochelle Park, NJ: Hayden, 1978), v.

14. Wynton Marsalis, *To a Young Jazz Musician: Letters from the Road* (New York: Random House, 2004), 11.

15. "CS9: Problem-Solving for the CS Technical Interview," accessed January 14, 2018, http://web.stanford.edu/class/cs9/.

16. Manfred Broy and Ernst Denert, eds., *Software Pioneers: Contributions to Software Engineering* (Berlin: Springer Verlag, 2002).

17. National Public Radio, "How One College Is Closing the Computer Science Gender Gap," accessed January 14, 2018, http://www.npr.org/sections/alltechconsid ered/2013/05/01/178810710/How-One-College-Is-Closing-The-Tech-Gender-Gap.

18. Lafayette College, "CS—Computer Science," accessed January 14, 2018, http://catalog.lafayette.edu/en/current/Catalog/Courses/CS-Computer-Science. This is just one example of this.

19. Computing Research Association, *Generation CS: Computer Science Undergraduate Enrollments Surge since 2006* 2017, B-1, accessed January 14, 2018, http://cra.org/data/Generation-CS/.

20. Ibid., D-2.

21. US Census Bureau, "Overview of Race and Hispanic Origin: 2000," March 2011, 4, accessed January 14, 2018, https://www.census.gov/prod/cen2010/briefs/c2010br-02.pdf. The underrepresented minorities are blacks/African Americans, Hispanics/Latinos, American Indians, Alaska natives, and native Hawaiians or other Pacific Islanders.

22. Harlan D. Mills, "Software Engineering Education," in *Software Productivity* (New York: Dorset House, 1988), 260.

23. Frederick P. Brooks Jr., "The Tar Pit," in *The Mythical Man-Month: Essays on Software Engineering*, anniversary ed. (Boston: Addison-Wesley, 1995), 9.

24. Peter Seibel, *Coders at Work: Reflections on the Craft of Programming* (New York: Apress, 2009), 468.

25. Henry Baird, interview with the author, August 7, 2017.

26. David Parnas, e-mail with the author, May 18, 2017.

27. Donald E. Knuth, "Literate Programming," *The Computer Journal* 27, no. 2 (January 1984): 97.

28. Ibid.

29. Donald Knuth, phone interview with the author, February 10, 2017.

30. Matt Pharr and Greg Humphreys, *Physically Based Rendering: From Theory to Implementation* (Amsterdam: Morgan Kaufmann, 2004).

31. Christopher Fraser and David Hanson, *A Retargetable C Compiler: Design and Implementation* (Redwood City, CA: Benjamin-Cummins Publishing, 1995).

32. Pharr and Humphreys, *Physically Based Rendering*, 1.

33. Knuth, "Literate Programming," 97.

34. Pat Hanrahan, foreword to *Physically Based Rendering: From Theory to Implementation*, by Matt Pharr and Greg Humphreys (Amsterdam: Morgan Kaufmann, 2004), xxii.

35. Actually not every one; there are certain exceptions, such as dividing by zero or the stack running out of memory, which are so unexpected—and also can happen anywhere, essentially—that they don't need to be listed.

36. David Parnas, Jan Madey, and Michal Iglewski, "Precise Documentation of Well-Structured Programs," *IEEE Transactions on Software Engineering* 20, no. 12 (December 1994): 948–976.

37. Steve McConnell, *After the Gold Rush: Creating a True Profession of Software Engineering* (Redmond, WA: Microsoft Press, 1999), 39.

38. Mills, "Software Engineering Education," 255–256.

39. Frederick P. Brooks Jr., *The Mythical Man-Month: Essays on Software Engineering*, anniversary ed. (Boston: Addison-Wesley, 1995), viii.

40. Andy Oram and Greg Wilson, eds., *Making Software: What Really Works, and Why We Believe It* (Sebastopol, CA: O'Reilly, 2011).

41. Robert L. Glass, "How Can Computer Science Truly Become a Science, and Software Engineering Truly Become Engineering," in *Software Conflict 2.0: The Art and Science of Software Engineering* (Atlanta: Developer.* Books, 2006), 231.

42. McConnell, *After the Gold Rush*, 84–88, 101–112.

43. Ivar Jacobson, Pan-Wei Ng, Paul E. McMahon, Ian Spence, and Svante Lidman, *The Essence of Software Engineering: Applying the SEMAT Kernel* (Upper Saddle River, NJ: Addison-Wesley, 2013), 255.

44. Texas Board of Professional Engineers, "Software Engineering," accessed January 14, 2018, https://engineers.texas.gov/software.html.

45. National Council of Examiners for Engineering and Surveying, "NCEES Principles and Practice of Engineering Examination Software Engineering Exam Specifications," April 2013, accessed January 14, 2018, https://engineers.texas.gov/downloads/ncees_PESoftware_2013.pdf.

46. David Parnas, e-mail to the author, May 22, 2017.

47. McConnell, *After the Gold Rush*, 103.

48. Ibid., 155.

49. Allen Ginsberg, "Howl," in *Howl and Other Poems* (San Francisco: City Lights Books, 1956), 9.

50. Steve Lohr, *Go To: The Story of the Math Majors, Bridge Players, Engineers, Chess Wizards, Maverick Scientists, and Iconoclasts—the Programmers Who Created the Software Revolution* (New York: Basic Books, 2001), 99.

51. Kurt Anderson, *Fantasyland: How America Went Haywire* (New York: Random House, 2017), 174.

52. Mills, "Software Engineering Education," 253.

53. Edward Yourdon, *Decline and Fall of the American Programmer* (Englewood Cliffs, NJ: Yourdon Press, 1992), 1, 17, 20.

54. Edward Yourdon, *Rise and Resurrection of the American Programmer* (Upper Saddle River, NJ: Yourdon Press, 1996), xi.

55. *Fast Times at Ridgemont High*, directed by Amy Heckerling (1982; Universal City, CA: Universal Studios Home Entertainment, 2004), DVD.

Index